THE DUBUQUE SHOT TOWER

EDITED BY JOHN ADELMANN

THE History PRESS

Published by The History Press
Charleston, SC 29403
www.historypress.net

Back cover, middle: Ronald Tigges, Digitaldubuque.com.

First published 2011

Manufactured in the United States
ISBN 978.1.60949.256.4

Library of Congress Cataloging-in-Publication Data
The Dubuque Shot Tower / edited by John Adelmann.
p. cm.
ISBN 978-1-60949-256-4
1. Shot Tower (Dubuque, Iowa)--History. 2. Lead industry and trade--Iowa--Dubuque--History. 3. Dubuque (Iowa)--History. I. Adelmann, John.
F629.D8D834 2011
977.7'39--dc23
2011018250

Notice: The information in this book is true and complete to the best of our knowledge. It is offered without guarantee on the part of the author or The History Press. The author and The History Press disclaim all liability in connection with the use of this book.

Contents

Contents

Preface

Lead shot production at the Shot Tower built by George W. Rogers & Co. in Dubuque, Iowa, spanned about six years, from December 1856, when it opened for business, to July 1862, when it was boarded up and sold to a rival shot-making firm in St. Louis. However, to uncover the history that preceded the Shot Tower's existence, to learn who built, owned and operated it, and then chronicle the events that took place after its shot-making days were done would rival the most rigorous academic syllabus. This in-depth research project blended the disciplines of social studies and English language arts. It focused on learning how to access and interpret primary sources and writing for an audience and emphasized the making of community connections in order to tell the story.

"What is shot?" "Why is this thing so tall?" "What did they *do* here?" were the superficial questions the American history and English students at Central Alternative High School initially asked when they were presented with the general focus of the class—to uncover the history of the Shot Tower. However, their discussions soon led to more sophisticated, open-ended inquiries and statements that reflected a desire to reach beyond the need to answer questions on an exam: "What does the Shot Tower stand for today?" "We need to tell this story so people will appreciate what the Shot Tower means to our community." This developing attitude of discovery kept the budding researchers focused and engaged. The satisfaction they derived from uncovering the Shot Tower's history spurred them to seek more connections and share what they learned with others. It was a sight to behold.

The original goal of the Shot Tower class in 2005–6 was to complete the research and then publish a book that would tell the story of our community's famous landmark to coincide with its 150th anniversary and the city's rehabilitation project. However, whenever students are encouraged to take an active role in charting the direction of their class, teachers can find themselves in unexpected situations. One day, a particularly gregarious student suggested that we create a *calendar* instead of a book to celebrate the Shot Tower, and a spirited discussion ensued. Eventually, by a show of hands, the class embraced his idea. The group's enthusiasm was then redirected toward achieving this exciting new goal. The concept was presented to businesses and civic groups, and the young scholars helped raise funds to publish "The George W. Rogers & Co. Shot Tower Commemorative Calendar." It went on to earn the prestigious Loren Horton Community History Award, presented by the State Historical Society of Iowa, and the Preservation at Its Best Award, presented by the Iowa Historical Preservation Alliance in 2007. Even though much of the research did not fit the calendar format, there was always an expectation that a book about the Shot Tower eventually would become a reality. That day has now come.

I am grateful to several individuals who helped the students create this Shot Tower research book. Central English teacher Tim Ebeling, whose passion for the written word is surpassed only by his determination to see every student succeed, was an invaluable colleague and collaborator and helped guide the students to share with the community what they were learning as he cultivated their writing and public speaking skills. Des Moines historian Jim Jacobsen's groundbreaking study of the Shot Tower helped the City of Dubuque secure federal and state grants to underwrite the iconic structure's rehabilitation. His findings streamlined our searches through microfilm and primary source documents. Mike Gibson, director of the Center for Dubuque History, welcomed our students to Loras College, invited us to make his extensive repository our field classroom and helped the students conduct research by providing them rare sources of information to complete their study. Tacie Campbell and Denise Vondran, the respective curator and registrar of the National Mississippi River Museum and Aquarium, along with registrar Jim Wall-Wild, also opened their facilities to our students, making available additional material on local history. Archaeologist Mike Gregory unearthed important new information about the Shot Tower's interior construction in the early days of the rehabilitation. Engineer Ron Balmer's interest in restoring the Shot Tower motivated him to produce a fascinating PowerPoint presentation, information which he graciously shared with us and energized our students. Laura Carstens, planning services manager for the City of Dubuque, kept us current with the progress of the Shot Tower's rehab project and provided important chronological information. Gerda Preston Hartman, a dedicated and respected historian, volunteered her time and expertise in promoting the book and helped the students uncover material on Dubuque's early history and the Mines of Spain in the years prior to the city's incorporation and the Shot Tower's construction. Central's paraprofessional Carol Cross helped students polish their drafts and tracked down historical photos and images to enhance their work. Dave Simon volunteered as a project mentor and gave our students another perspective of local history inside and outside the classroom. Historian Bill Hammel shared information from his extensive files on Dubuque fire department history and on the life of Shot Tower owner J.K. Graves. Longtime Central supporter Norma Denlinger worked behind the scenes proofreading chapters and developed timely community support. Ben Gibson of The History Press shepherded us through the publishing process, making sure this book turned out the best it could be. I also thank my wife, Judy, for her faithful patience and computer wisdom, and I acknowledge the Master Teacher, without Whose guidance this project never could have succeeded.

During this academic journey into our local history, student researchers made life-changing connections with people and events of days gone by. They grew determined to tell the compelling story that they hoped would renew interest in that majestic structure built over 150 years ago that still stands a silent guard by the Mississippi River in Dubuque. Even though the Board of Education closed Central Alternative High School in 2010, the academic work that was completed in 2005 and 2006 will be honored and shared with the community as originally intended. While this may not be the definitive history of the Shot Tower, it certainly is the first of its kind. Slight overlapping of information from chapter to chapter may occur on occasion, but the intent was to preserve the originality and creativity of each student's contribution. May the story of the George W. Rogers & Co. Shot Tower add to the scholarship of Dubuque's fascinating history and help future generations appreciate the present by better understanding their past.

Chapter 1
Midwest Lead Deposits Influence Native American Life

By John Adelmann

There is an expression, "geography is destiny." Would the ancient Egyptians be remembered for their great civilization had it not been for the Nile River? Would Imperial Japan have set a course of expansion and conquest in the 1930s had it not been an island nation short of natural resources? Would Dubuque have become the oldest city in Iowa that boasts a shot tower without the presence of lead ore in the area?

Much has been written about Iowa's rolling hills of corn, soybean fields and livestock, but less is known about the mineral deposits that first attracted prehistoric Native American tribes and then early European explorers to this area of North America. In addition, while the personal and business relationships that developed between French voyageurs and the Native Americans who lived in this area are reasonably well documented, information about the mining activity of local tribes in the time before European contact is less complete.

James Dockal, professor of earth sciences at the University of North Carolina–Wilmington (and native Dubuquer), writing in "Lead and Zinc Mining in the Dubuque Area" in *Iowa Geology* 9 (1984), answers the question of why lead deposits are present here:

> *The zinc-lead deposits of the Dubuque area and the Upper Mississippi Valley (UMV) are part of an important class of metallic, hydrothermal*

> *(formed by hot water) ore deposits that are referred to collectively as Mississippi-Valley-Type (MVT) deposits. Hydrothermal deposits are known to be derived from hot vapors escaping from cooling igneous magmas, typically formed in mountain-building regions. MVT deposits, however, occur in areas of relatively flat-lying sedimentary rocks in the continental interior, distant from any possible igneous sources. The term "Mississippi-Valley-Type" deposit was coined because several of these ore fields form major mining districts in the drainage basin of the Mississippi River.*

Referring to research developed in 1839 as part of a "Report of a Geological Exploration of Part of Iowa, Wisconsin, and Illinois," archaeologist Larry Abbott explains how lead in the Upper Mississippi Valley appeared when it was discovered:

> *Much lead is raised from what are called "clay diggings." In these the ore is commonly found in detached fragments, but occasionally in regular sheets; and is covered up, sometimes to a very considerable depth, by clay, sand, or crumbled rock. The ore appears to have been dropped into its present position, in proportion as the rock which originally encased it gradually sunk under decaying influences...The veins (of ore) occasionally may still be traced horizontally for some distance along the ground where they may be supposed to have fallen, until they dip into a vertical fissure.*

Abbott adds that "float ore," a term used to describe mineral that is exposed at ground level or simply rests on the bedrock surface, can also be found as "channel lag" in creeks of this region. Not surprisingly, "float" galena was the first ore that was found and processed in the Mississippi Valley. Early Native American lead miners easily harvested this mineral found in bedrock fissures along the Mississippi River bluffs. Abbott also writes that much of the lead in this region "was derived from galena, a lead sulphide." Galena is 80 percent lead by weight, but a portion of this amount is lost in the commercial smelting process. Known for its gray color and streak, galena is brittle, fractures into cubed crystals and is found throughout the area of southwestern Wisconsin, northwestern Illinois and northeast Iowa.

Early Native American lead-mining activity in eastern Iowa began long before European explorers came on the scene. However, not much information about the techniques the Indians developed to extract the lead has been preserved. Historians have been forced to rely on the reports of

Europeans who made contacts with Native American miners and recorded what they saw. In a paper entitled "Prehistoric Lead Mining in the Dubuque Area," published in the *Journal of the Iowa Archeological Society*, researcher Donna Stubbs cites numerous studies that reveal evidence of what is called "post-contact" (AD 600–1800) Native American mining but adds that the documentation for earlier, "pre-contact" activity relies on secondary sources.

For example, we learn that the Sioux Indians worked lead mines in the Dubuque area, thanks mainly to the accounts provided by early French explorers in the late 1600s. Stubbs concludes that developing information on the early history of non-whites and their lead-mining efforts in the Upper Mississippi River Valley is difficult, in large part because of the ways the pre-contact Indian tribes extracted and used this mineral. She writes, "Prehistoric Native Americans did not smelt lead, but used it in un-smelted forms for a variety of functions. This, plus their primary procurement method of exploiting surface lead deposits, has left little direct evidence of prehistoric lead mining." In addition, Stubbs says that since early explorers to the Dubuque area knew of the Sioux Indians' lead-mining activity in the early 1600s, it can be hypothesized that earlier native peoples also had knowledge of the lead deposits. Given the heavy exploitation of lead during the post-contact period, however, the less complicated digging sites of these prehistoric tribes probably have been destroyed forever.

Even though there is very limited evidence of early Native American mining activity, some physical remains in Dubuque reveal that prehistoric lead mining did take place. Early descriptions of Indian mining activity are reflected in an archaeological discovery found in the lower Catfish Creek valley just south of present-day Dubuque by the Iowa Department of Transportation. Identified as site 13DB799, this area features "an obviously tool-dug excavation along a gash vein at the base of a palisade-like dolomite escarpment."

Archaeologist James Collins reports that there is solid evidence of fires placed along the base of the bedrock. He further explains, in research completed in 2004 entitled "A Meskwaki Mine in the Upper Mississippi River Lead District," that excavations into the bedrock at the 13DB799 site appear to have been made with hoes, axes and crowbars. He also reveals that evidence of using fire "for the intentional heating of the bedrock face immediately below the excavations…is typical of the Indian technique for heating and cracking rock," and he concludes that based on this evidence, the Catfish Creek mining site "is identified as Indian, and probably a Meskwaki mine."

Physical evidence of prehistoric lead extraction along Catfish Creek in Dubuque. Using hoes, axes, crowbars, and by repeatedly cracking bedrock with fire and water, the Meskwaki Indians removed ore from this escarpment. *James Collins, Iowa Office of the State Archaeologist.*

Archaeologist Stubbs reports that the Mines of Spain district contains numerous sites from the period when French explorer Julien Dubuque was the only European allowed to work the Meskwaki-controlled deposits. In addition, she says there are over 150 prehistoric sites in the Mines of Spain district, and over 200 prehistoric sites have been discovered and identified in the Dubuque vicinity. These sites include rock shelters, villages, cemeteries and ceremonial mounds.

According to Stubbs, the earliest evidence of lead from an Upper Mississippi River Valley source is found in sites during the Late Archaic–Early Woodland periods (3500–500 BC) from eastern Ontario to Louisiana. During the Middle Woodland period (200 BC–AD 500), the amount of galena lead ore that was exchanged and distributed by Native Americans was greatly increasing. Citing additional research, Stubbs explains that "the frequency of occurrence of large nodules and cleaved pieces of galena at Middle Woodland sites indicated that ores were not only collected at sources but that some mining was also conducted." She adds that artifacts found at archaeological sites reveal that Native Americans used lead in various ways. The ground-lead objects include bird effigies, ear spools, buttons, dice, beads, crystals, nodules, polished cubes and rectangles and ground cores.

Ethnographic sources indicate that lead was considered to have magical properties and was ground to a powder and used as face paint and for ceremonial uses. Larry Abbott reports that galena was sometimes carved by Indians "into various forms, powdered to be used as paint, or simply left natural." He adds, "It has been found archaeologically in both mortuary and habitation contexts," meaning that evidence of lead use has appeared in grave sites as well as in the remains of living quarters. In addition, John Walthall, in his 1981 research entitled "Galena and Aboriginal Trade in Eastern North America," offers an explanation for why galena is found in these sites. Quite simply, lead became more valuable in the eyes of the Indians if and when it came from a long distance away: "Based on data, a proposition may be generated: The greater the distance from the geological source area, the more valued a commodity (galena) becomes and the more likely it will be utilized in contexts of prestige and recognition behavior."

Walthall adds, "The proposition implies that if galena derives from a distant source then it should be treated as a valuable commodity and should occur in burial context with high-ranking individuals." Walthall also theorizes that the opposite is probably true when the lead source is found near an inhabited village or settlement. This galena would most likely be found in what he terms "non-burial, non-prestige-related contexts, such as domestic dwellings."

While there is considerable evidence in and around Dubuque of post-contact Native American mining activity, there is but scattered evidence that helps to reveal the Indians' mining techniques. Almost all historic references of Native American mining practices are provided by frontier traders of European descent. Collins cites an eyewitness account of Indian mining methods provided by explorer Henry Schoolcraft, based on a visit he made to the Meskwaki mines on Catfish Creek in the Mines of Spain region in 1820. Schoolcraft wrote:

Henry Schoolcraft (1793–1864). Noted geographer, ethnologist and early observer of Native American culture and mining activity, Schoolcraft is also credited with discovering the source of the Mississippi River in 1832. *Dubuque County Historical Society.*

> *The ore at these mines is now exclusively dug by the Indian women. Old and superannuated men also partake in the mining labor, but the warriors and men hold themselves above it. In this labor, the persons who engage in it employ the hoe, shovel, pick-axe, and crow-bar. These implements are supplied by the traders...who are purchasers of the crude ore. With these implements they dig trenches until they are arrested by solid rock. There are no shafts, even of the simplest kind, and the windlass and bucket are unknown to them—far more so the use of gunpowder in the mining operations. Their mode of going down into the deepest pits, and coming up from them, is by digging an inclined way, which permits the women to keep an erect position in walking.*

Collins cites another description of Native American mining activity provided by an Illinois miner in the 1820s. Moses Meeker's research, entitled "Early History of the Lead Region of Wisconsin," reveals specific details about the ways Indians separated ore from rock:

> *They would form an inclined plane where they went deep. I saw one place where they dug forty-five feet deep. Their manner of doing it was by drawing the mineral dirt and rock in what they called a "macock," a kind of basket made of birch bark. Their tools were a hoe made for the Indian trade, an axe, and a crowbar, made of an old gun barrel flattened at the breach, which they used for removing the rock. Their method of blasting was rather tedious, to be sure: they got dry wood, then kindled a fire along the rock as far as they wished to break it. After getting the rock hot, they poured cold water upon it which so cracked it that they could pry it up.*

Collins also points to research conducted by historian Bruce Mahan, who relied on French sources to describe an Indian mining excavation as follows: "Ledges containing ore were laboriously cracked by building a fire beneath, then pouring cold water on the heated rocks."

During the 1600s and early 1700s, French voyageurs and the Native Americans who smelted lead became "the first metallurgists in the upper Mississippi valley," according to Abbott. However, he says that the Indians' ability to develop their processed lead production was limited because their refining techniques were pretty basic. Fires often were made in old tree stumps, into which the chunks of ore were deposited. When the fires died down, clumps of lead that had been rendered from the ore were retrieved. This rather haphazard operation was sufficient enough to meet the personal needs of the traders and Indians and, to a lesser degree, fulfilled the demand for trading purposes. Abbott cites the research done in 1980 by Carl Sauer, whose work entitled "Seventeenth Century North America" explains the primitive process both the French explorers and the natives used to smelt lead ore:

> [The log hearth] *was simply a trench dug into the hillside that was filled with logs on which the ore was thrown. The fire would be started in the evening, and in the morning the lead would be found at the bottom of the trench, where it had consolidated into shapeless lumps or had trickled into small holes that had been scratched in the earth under the logs.*

Sauer then explains an ingenious technique that was developed to create lead ingots so they could be more easily transported:

> *Later, these frontier metallurgists found it convenient when casting their lead to make a hole in the ground in the shape of a brick, and in the center*

> *of this mold they placed a stick, so that when the lead ran into the hole from the ore that had been reduced by a wood fire, it cooled to an ingot with a perforation in it; through this perforation they drew a rawhide rope, to facilitate transport, for it could be then easily swung either on the shoulder of a man or on the back of a horse.*

After the technique of smelting lead ore was introduced to the Native Americans in the post-contact period, researcher Stubbs says that some of this processed lead was eventually used as decorative inlays in Native American smoking pipes and other objects.

Despite the findings at the Mines of Spain and evidence of more recent mining activity by local Indian tribes, researcher Stubbs makes the following conclusion in 2004 about earlier, pre-contact Indian lead mining in this area:

> *To date, the only evidence for prehistoric lead mining in Dubuque is circumstantial and indirect. Predictive models and hypotheses suggest that the best possibility to confirm prehistoric lead exploration in Dubuque is in a habitation or village in close proximity to naturally occurring lead deposits. During the Southwest Arterial Highway project, one archaeological site, 13DB707, was identified that meets these two criteria. Future investigations at this deeply buried site may confirm whether Native Americans exploited the lead deposits in the Dubuque area.*

The discovery of lead ore by Native Americans in the vicinity of Dubuque set the stage for future mining by European traders who ventured into the Midwest in the late 1600s and early 1700s. Lead, the simple, natural product that was used for religious and personal applications by the Indians, would eventually attract prospectors who saw its full-scale mining as an opportunity to attain wealth and prosperity.

Chapter 2
Miami Tribe Leads French Explorers to Lead

"Focus on Youth" essay by Maranda Duhme

Editor's note: Central Alternative High School students in Dubuque, Iowa, were invited to submit written essays to promote their Shot Tower research project in the Dubuque Telegraph Herald*'s weekly "Focus on Youth" column. Twenty-four articles were published between 2005 and 2006, seventeen of which are included in this book. The remaining submissions were condensed versions of longer research papers that now appear here as chapters.*

For research, I read a paper written by James Collins, from the Iowa Office of the State Archeologist, and discovered that the present-day Mines of Spain were once labeled the Mines of Nicholas Perrot on maps and in journals of a European explorer named Pierre Charles Le Sueur. Collins says, "Knowledge of the mine's existence was widely held on this continent and in Europe." So who was Nicholas Perrot?

Perrot served in the French military and was stationed at a trading post at Prairie du Chien in 1685. In 1690, he received a piece of lead ore, also known as galena, from a delegation of Miami Indians who lived some fifty miles downriver. Collins states that galena was highly prized by many American Indians who carried pieces of it as "magical charms." They used it in powdered form to make face paint and in their religious ceremonies. Perrot eventually found that the source of the galena was at the mouth of Catfish Creek. At this time, most of the area's lead was mined by Indians,

This map shows the location of Julien Dubuque's mines in 1804. Shaded land area was ceded by the Meskwaki tribe. The territory west of the Mississippi River would later become Iowa. *John Wilkie,* Dubuque on the Mississippi, *Loras College Press.*

but Frenchman Jean Marie Cardinal and a few others of European descent were also working the fields.

The Meskwaki tribe settled in Iowa in the 1730s, and by 1780, they had a permanent village at the mouth of Catfish Creek, just before Chief Peosta's wife discovered a rich deposit. Within a few years, French trader Julien Dubuque arrived on the scene and ingratiated himself to the Meskwaki and eventually married into their tribe. In 1788, he got written permission to operate and work their mines, but he could not own them. Ore was taken to his trading house on the western bank of the Mississippi. Eventually, Dubuque became more recognized for his lead-mining activities than Perrot.

During this time, Spain controlled much of the North American Midwest, including the Iowa Territory. Dubuque must have been quite a character because he decided to flatter the Spanish rulers and called his operation the "Mines of Spain." Jim Collins says in 1796, Dubuque "humbly petitioned" the Spanish governor of Louisiana for "formal recognition" of his mineral rights and got a parcel of land "extending twenty-one miles along the Mississippi and nine miles inland roughly centered on what is now the city of Dubuque."

Julien Dubuque worked his mines but eventually wound up heavily in debt and died in 1810. He is buried on the bluff that overlooks Catfish Creek, which is now an important place in the mining history of Dubuque and Iowa.

Chapter 3

Bringing History to Life

"Focus on Youth" essay by Tawni Bauer

Central Alternative High School's Shot Tower research effort isn't the typical high school project. There isn't much documented about the Shot Tower in regular history textbooks, so we have to locate all of the information ourselves. But we cannot go to the library and expect there to be endless numbers of books just about the Shot Tower. However, there is information scattered around. One needs the ambition to find it.

To get our information, we have to dig at least 150 years back into the history of Dubuque. Presently, I am working on a chapter about the Mines of Spain. Without these lead mines, there would have been no need for the Shot Tower. Without the Shot Tower, Dubuque's history would be different.

I am getting a lot of my information from nomination papers written by Gerda Preston Hartman that she submitted to the National Register of Historic Places, and from Gerda herself. She is a local expert and a historian who really knows her facts. Hers was the second family to come to Iowa to settle. I attended a presentation she gave at Dubuque's Carnegie-Stout Library to hear her talk about her family.

Gerda told us about "Swedish Annie," the title of a book she wrote about her grandmother, who came to Iowa from Sweden in 1900. Annie had a relationship with a man her father didn't approve of. Gerda found the love letters from Annie to her paramour. The book tells the story behind those letters.

Gerda also talked about the Mines of Spain. She didn't have much time the evening of the presentation to do that, but what she did tell us sparked my interest in the mines and their importance to the story of the Shot Tower. I plan on having many meetings with Gerda so I can write a chapter for the Shot Tower research book. There is no better person to get information from. She and her family own land on the Mines of Spain to this day that was passed down from her ancestors who came to Iowa in the 1830s.

Gerda also has a museum at her home that contains many family heirlooms and personal artifacts, including some of Swedish Annie's belongings. At the presentation at the library, she was wearing Annie's bonnet and jewelry. She also had a case full of other belongings of Annie's and some artifacts from the miners on display. It was very exciting to see these things and think about how they were valued by people over one hundred years ago.

During my time in the Shot Tower class, I realized that it is possible to learn about history from real people instead of just textbooks. The Shot Tower is an important part of our local history, and with Gerda's help, we will be able to teach others about it.

Chapter 4
The Mines of Spain

By Tawni Bauer

Editor's note: The following information was adapted from documents prepared by historian Gerda Preston Hartman that were submitted to the National Register of Historic Places to recognize the Mines of Spain's significance in Dubuque's and our nation's history.

Mining activity at the Mines of Spain that involved white explorers and settlers occurred from about 1788 to 1865. This was part of the first major mining rush in the United States. Lead mining served as a major reason for the settlement of Illinois, Iowa and Wisconsin. Julien Dubuque held an important role in this. He is the one who actually named the mines. Dubuque bought the right to work the land from the Indians. However, he did not actually own the land; he just purchased the right to mine it. There was also a set of laws, rules and customs that was developed that kept the peace among the lead miners and protected their claims. This was the first system of self-government to regulate mining activity in America, and the miners who left Iowa for California during the gold rush of 1848–49 used these regulations to help protect their claims in the West. This is why the Mines of Spain has local and national significance.

Although Julien Dubuque is the most popular name associated with the Mines of Spain, he actually did not discover the lead deposits. The Miami Indians found them and told French explorers about them. In 1685, Nicholas Perrot (1644–1717) arrived in Prairie du Chien and built a fort near where the Wisconsin River joins the Mississippi. He learned to speak several Indian languages and gained their trust by trading with them. In 1690, Perrot was invited by representatives of the Miami Indian nation to visit their settlement about fifty miles south of his fort, in hopes that they could trade furs with him. Some historic accounts say that Perrot was given a hunk of shiny mineral by one of the Miami chiefs as an afterthought, but he instantly recognized it as lead ore. No other explorer had found lead deposits in the New World, but Perrot knew the value of this gift. He then agreed to set up another fort across from the site where the Miami found the lead. Today, this spot is called Gramercy Park in East Dubuque, Illinois.

For some reason, Perrot left the area in 1696, most likely because his trading license was not renewed by France. Then another Frenchman, Pierre Charles Le Sueur, tried to paddle the entire Mississippi River. While on this journey, he and his men stopped at Perrot's mines in August 1700. Le Sueur took some of the lead he found there back to France in 1702, and a year later, the first maps of what was then called "New France" had Perrot's mines pinpointed on them. However, only the Indians mined lead in this area after Perrot left. Eventually, the Miami tribe abandoned their mines and river settlements and moved east into the territory of Illinois.

In the 1730s, the Meskwaki tribe entered the region and began to mine lead. By the 1780s, they had set up a village in the area where Catfish Creek meets the Mississippi River just south of present-day Dubuque. Because there was success at these mines, news spread through the Upper Mississippi River Valley, and the Indians continued to trade with Europeans who traveled the river. One of those men was Julien Dubuque, and he wanted to mine lead with them.

Dubuque was born on January 10, 1762, in Quebec, which at the time was French territory. By 1785, he was living in Prairie du Chien. He heard of the lead deposits in the Meskwaki territory and took an instant interest in them. On September 22, 1788, he came to an agreement with the Meskwaki, who allowed him to work the land for lead. He built a trading post, a residence and a smelter for purifying the lead at the mouth of Catfish Creek on the Mississippi River. With help from his French Canadian friends, he was able to open the mines. In 1796, Spain now ruled this part of the New World, and the Spanish governor finally recognized Dubuque's rights to mine the land. He tried to flatter the governor by naming the territory the "Mines of Spain."

At Dubuque's post, he combined fur and lead trading and basic smelting of the lead ore. This trading of lead provided the Meskwaki with about

Mural painted by Chicago artist James McBurney depicts Julien Dubuque inspecting lead samples as local Fox chiefs look on. At left, Indian women emerge from the mine, hauling loads of ore on their backs. *Center for Dubuque History.*

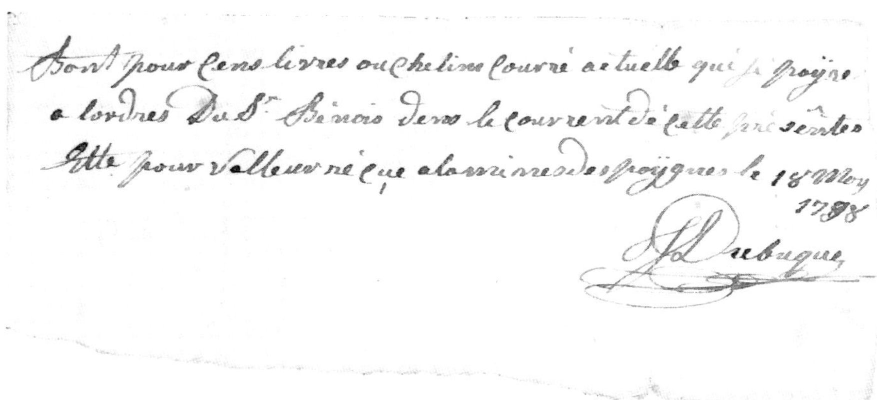

Bont pour Cens livres ou Chelins Courré actuelle que je payeré a lordres Du Sr. Binois dans le courrent de cette présentes ette pour valleur reçue a la mines des Sayques le 18 May 1798

J. Dubuque

A promissory note written in French "good for 100 livres which I will pay to the order of St. Benois…for value received…for arrears," signed by Julien Dubuque on May 18, 1798. *Kenneth W. Rendell Gallery.*

half of their needs in goods. Each year, Dubuque also traveled down the Mississippi River to St. Louis. He loaded up his boats with furs and lead. When he arrived, he traded for other goods. Dubuque loved the shops there, and it seems he spent too much. Due to this, he fell into debt. To get himself out of this problem, he assigned about half of his claim of the mines to

Auguste Chouteau, a man who had loaned him the funds to purchase the rights to mine the land in the first place. This "deal" would become a real problem for Iowa miners and settlers later on when Chouteau tried to claim this land as his own. The U.S. Supreme Court eventually ruled in 1853 that he had no rights to the land because his friend Julien Dubuque didn't have rights to the land either.

Julien Dubuque died on March 24, 1810. To show him respect and gratitude, the Meskwaki buried him with tribal honors in a log tomb on the bluff overlooking where Catfish Creek enters the Mississippi River. In 1897, Dubuque's body was dug up and reburied under a new monument. To this day, the limestone grave still stands on the bluff overlooking the Mississippi River and Catfish Creek. The Meskwaki learned a lot from Dubuque, including how to mine and smelt the lead ore.

Many other mines were discovered, and smelters were built in towns such as Mineral Point, Wisconsin; and Hardscrabble, Swindler's Ridge, Buncombe Diggins and Galena, Illinois. While all this activity was happening, the Meskwaki continued to mine their land under the protection of a group of

The skeletal remains of lead miner Julien Dubuque, minus fingers, toes and a jawbone that was reportedly stolen, just before being reinterred in the limestone monument on the bluff overlooking Catfish Creek in 1897. *Dubuque County Historical Society.*

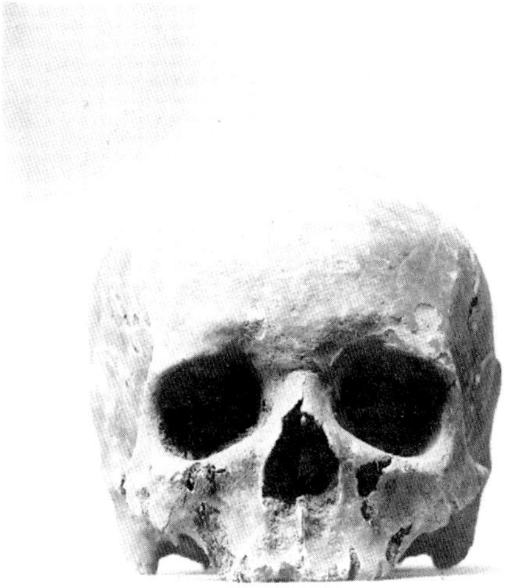

The skull of Julien Dubuque. *Dubuque County Historical Society.*

soldiers led by Colonel Zachary Taylor. The Mississippi River acted as a natural boundary for the Indians, and the U.S. government tried to shield them from impatient white miners. Taylor's post was located at Fort Crawford, a spot about fifty miles up the Mississippi from Prairie du Chien. However, even with this military force, there was a rush of miners to the Meskwaki lands. At one point, the Meskwaki left their lodges to go to Prairie du Chien on a peace mission. While they were gone, the Americans slipped in and tried to claim the mines for themselves. However, Colonel Taylor followed his orders and sent troops in to evict the intruders. The Meskwaki were not given peace, however.

In 1830–31, the Indians had a very hard winter. They were short on food and were being threatened by some miners in the surrounding area. That spring, the Meskwaki left and never returned. At Rock Island, they met with Sauk chief Makataimeshekiakiak, who was also known as Black Hawk, and agreed to join him in a war that they hoped would push the white settlers off their lands. However, the fighting in 1832 greatly reduced the number of the Sauk and Meskwaki tribes. Those who were not involved in the war were removed from their land by the new treaty that was signed with the U.S. government when the war ended. One historian stated, "The treatment of Black Hawk and his band, both before and during the hostilities is discreditable to us. It is a black chapter in the history of the West."

Once the Meskwaki were removed, miners crossed the river and rushed to claim the ore in the Dubuque mining area that extended twenty-one miles along the western bank of the Mississippi River. Soon, individuals from throughout the country were attracted here because of newspaper articles like the one printed in the Wheeling, Virginia *Gazette*, which stated, "Go to that land for we are confident there you will find the Eldorado of your imagination. Are you poor? Seek, and you shall obtain a competence. Art thou rich? There shall your monies reap a noble interest."

George Wallace Jones (1804–1896). A non-voting delegate who represented the Michigan Territory in Congress from 1834 to 1837, Jones later was elected as one of the first two senators from the new state of Iowa in 1848. *Dubuque County Historical Society.*

During the mid-1830s, a major settlement in the lead-mining district of eastern Iowa took root, and it had an impact on the governments of territorial Wisconsin, established in 1836, along with the Iowa Territory, which was established in 1838. The first delegate to Congress from this region was George Wallace Jones, who was among the most important of the early miners. He was the son of a very wealthy Missouri man who owned mines and smelters. Jones arrived in the region in 1827 and leased 1,001 acres in southwest Wisconsin, across the Mississippi River from Dubuque. He also operated a smelter and a post where he traded with the Meskwaki. Jones was an officer in the militia during the Black Hawk War and bought and sold land throughout the lead district, including the Dubuque mines. He was obviously an important individual because he was selected as one of the first senators from Iowa when it became a state.

The 1840s brought even more miners to this area of the country. The ore that was discovered and raised from the deeper mines produced better-quality lead and was found in larger amounts. These factors contributed to the building of the settlement of Dubuque and eventually led to the construction of the George W. Rogers & Co. Shot Tower, where the process of making shot would imitate the method discovered in England in the late 1700s.

Chapter 5
The World's First Shot Tower

By John Adelmann

From the early days of firearms manufacturing in the 1200s, makers of ammunition struggled with how to produce good-quality shot in large enough amounts to satisfy the demand. Some shot was made by pouring molten lead into rounded molds, but the results often left irregular seams, dimples and other marks on the shot surface, lessening its quality and aerodynamic efficiency. This method was also very time-consuming. Other processes were tried, including pouring molten lead through sieves that were suspended several inches above a tub of water. However, this usually resulted in creating egg-shaped shot with long, unsightly tails. As the demand for military and sport ammunition continued to grow, arms makers searched for new and more effective ways of making shot. According to Professor Walter Minchinton, author of "The Shot Tower," published in *Invention & Technology* (Spring/Summer 1990), a process patented by an Englishman in 1782 revolutionized the shot-making industry. This observant man would eventually transform his humble home into the world's first shot tower.

William Watts was a plumber who lived in Bristol, a city that was known as the center of English lead manufacturing during the late 1700s. The shot-making industry in Britain was well established by this time, and Watts probably was familiar with the process of making small shot but decided to experiment with the height of the drop to make larger, rounded pellets. There are several versions of how he came to develop the process for which

Producing handmade lead ammunition on the frontier was tedious and inefficient. The method shown here created surface irregularities and imperfections on the molded spheres that affected shooting accuracy before large-scale shot manufacturing, sorting and finishing became common. *Janice Myelle.*

he would eventually take out a patent. According to information taken from www.bristoltours.com, one account says Watts dreamed that he was caught up in a heavy rainstorm, but instead of raindrops falling from the sky, "there came down molten drops of lead, which, to his intense surprise, fell in a perfectly globular form."

If the legend is to be believed, Watts was haunted by this idea and decided to see if molten lead indeed would perform as it did in his dream. He climbed the tower of St. Mary Redcliffe Church in Bristol and dropped some liquid lead into a tub of water he positioned below (presumably with the permission of the local pastor!). The experiment was a success. He then quickly built a shot tower at the top of Redcliffe Hill right in front of the church and began producing shot.

Another version of Watts's "discovery" gives more credit to his wife, who is said to have seen in a dream some molten lead "fall from the top of a spiral staircase into a vessel of water beneath, and she was perceptive enough to notice that it assumed a perfectly spherical shape." Her attempts to tell her husband about the "process" she dreamed about fell on deaf ears. After dreaming the same dream once again, Mrs. Watts insisted that her husband listen to her. Watts then "began to think there might be something in it." He conducted experiments, and the rest became part of shot-making history. Minchinton claims that yet another legend of how shot-making was perfected simply has Watts being struck with inspiration when "he noticed that raindrops formed perfect spheres as they fell." Whatever story is the truth, Minchinton declares, "the idea he had is still the best we have."

The key to Watts's success came when he realized that lead had to be dropped from a great height, not just inches, from the cooling water in order to form into perfectly round pellets. His patent specified that the sieve that broke the molten lead into droplets had to be set at least 10 feet above the water and that 150 feet or more was necessary to produce the largest shot. In 1782, he took out a patent for his shot-making process that is defined in a book written by Helen Reid, titled *Bristol & Co.*, published in 1987: "A method of making small shot solid throughout, perfectly globular in form, and without dimples, scratches and imperfections which other shot, heretofore manufactured, usually have on their surface." With patent in hand, William Watts constructed three additional stories atop the house he lived in. According to Minchinton, Watts then cut holes in the floors of the house and dug a well beneath it to reach his desired 150-foot drop length. The author adds that Watts's tower "was an immediate success and made him rich."

Word of Watts's patented shot-making process spread rapidly, and his exploits were soon recorded for history in a clumsy poem written by a certain John Dix of Bristol and published in the *Bristol Mirror*:

Mr. Watts very soon a patent got
So that only himself could make Patent Shot.
And King George and his son declared they'd not
Shoot with anything else—and ordered a lot.
The Regent swore that the smallest spot
In a bird's eye he'd surely dot;
And every sportsman, both sober and sot,
From the peer in his hall to the hind in his cot,
Vowed that they cared not a single jot,
When the game was strong and the chase was hot,
For anything else than Patent Shot.

In 1787, William Watts took on a partner. The two men began to make enough money so that Watts eventually was tempted to sell his share of the shot-making business and turn to land speculation. He went bankrupt while in the process of building the foundations for the Windsor Terrace in Clifton, England, during the economic slump that was caused by the Napoleonic Wars. Never one to forget a favor, however, Watts named the terrace in honor of King George III, who had recognized his shot-making design.

Lead shot continued to be manufactured in Watts's original building with his patented process until 1968, when the structure gave way to English road modernization. Author Reid explained what happened: "In 1968 the road going up Redcliffe Hill was doubled in size and this led to the demolition of the Shot Tower. Another tower was built in Cheese Lane and the process continued for about another twenty years. Production ceased in 1995."

University of Houston professor John W. Lienhard observes in the website uh.edu/engines/epi422.htm that William Watts's development of the shot-making process reveals two important elements of what he characterizes as a "good invention":

> *The first is perception. Watts gazed more closely at nature and saw what other people had missed. The other element is simplicity. Others had labored to control the process with their own hands. Watts had the grace to stand aside and let nature do the work for him. The real beauty of this process is that, in the end, there is no human process at all.*

Professor Minchinton summarizes his study of the shot-making process by saying that while the technological development of some industries come about through gradual change over years of time, improvement in shot production was the result of one giant leap, thanks largely to Bristol plumber William Watts:

> *The improvements in shot tower design since the 1700s have been largely peripheral: an elevator to transport workers and lead to the top of the tower, different fuels to melt the lead, and in some cases a controlled mechanical feed to the sieve. The essentials of the 1782 patent design remain unchanged. In the history of technology, William Watts deserves to be recognized for bringing to the problem of shot-making a simple solution that has never been excelled.*

Even though William Watts patented his manufacturing process, conflict between England and France in the early 1800s forced the new American government to impose restrictions on imported English goods, including shot. This decision would drive American entrepreneurs to imitate the methods of William Watts to meet the demand for shot that was growing in their own country.

Chapter 6

Embargo of 1807 Awakens the Shot-Making Industry in America

"Focus on Youth" essay by Spencer David

William Watts was a plumber who lived in Bristol, England, in the 1780s. According to information we found, his idea of developing a shot-making process came to him while he watched falling raindrops! He realized that as droplets fell from the sky, they would turn into perfect spheres. Watts saw the possibility that if molten lead fell far enough through the air that it could also form into perfectly round balls.

Watts then went back to his house, which was already three stories high, and began adding on floors. After he completed the addition, it was twice as high as before. He put windows on each floor and installed a tub of water at the bottom to cool down the falling hot lead. He came up with a method to pour melted lead into a sieve and made it fall down 150 feet into the water. William Watts had created the world's first shot tower.

Watts's process greatly reduced the cost and labor for making high-quality shot. Soon, shot towers were appearing all over Europe, as well as in England. Some of this shot was imported by American gunsmiths. However, England and France were at war with each other in the early 1800s, and America was trading with both of them. Neither side liked the fact that we were trading with their enemy. When France and Britain announced they would capture any American ships trading with the other side, U.S. naval vessels and sailors were put in danger.

President Thomas Jefferson faced a great problem. If we kept trading with either England or France, America might get dragged into war. Jefferson then made a tough decision. In 1807, he asked Congress to place an embargo on all U.S. imports. This made it illegal for Americans to trade with foreign countries. His efforts wound up punishing many citizens in our country because they depended on foreign goods, including shot made in England. With no way to purchase foreign-made shot, American businessmen saw an opportunity to imitate William Watts's idea and make their own shot in American shot towers. Soon, these tall towers began to spring up across our nation. Today, there are only a few of these structures still standing. They can be found in Virginia, Connecticut, Maryland, Wisconsin, Pennsylvania and, of course, Dubuque, Iowa!

Chapter 7

Early Lead Traffic on the Upper Mississippi River Valley

By John Adelmann

Writing in the *Mississippi Valley Historical Review* 17, no. 1 (June 1930), historian William J. Petersen states, "No other single factor during the quarter century ending in 1848 was so important in developing steam-boating on the upper Mississippi as the shipment of lead." The combined factors of discovering the ore and the steamboat industry that transported prospectors to the lead fields, imported the provisions they lived on and exported the lead ore they mined would have a lasting effect on the development of the region. Naturally occurring lead ore was discovered in large quantities in northwestern Illinois, southwestern Wisconsin and northeast Iowa where it meets those other two states. In particular, the counties of Jo Daviess and Carroll in Illinois; Grant, Iowa and Lafayette in Wisconsin; and Dubuque in Iowa became the focus of miners and land speculators as word of the gray mineral found on the surface or just beneath the soil spread throughout the expanding young nation.

However, if the personal writings of the famed western explorer Zebulon Pike were to be consulted, the above glowing report of lead mining in the upper Mississippi River Valley, particularly in the Dubuque area, might have been challenged. The reason? In 1805, French explorer and miner Julien Dubuque met up with Pike during his visit to this area and told him that he had "only" manufactured between twenty thousand to forty thousand pounds of lead, which he regularly shipped down the Mississippi River to

American explorer Zebulon Pike (1779–1813). While asking questions to determine ore yields at the Mines of Spain during an 1805 visit, Pike observed that Julien Dubuque's answers "seem to convey with them the semblance of equivocation." *Dubuque County Historical Society.*

St. Louis. Historians now conclude that Dubuque was trying to keep his ore discoveries to himself and did his best to minimize the risk of other miners relocating and extracting lead. Research also shows that Pike sensed Dubuque's deception. He wrote that Dubuque's answers to his questions "seem to convey with them the semblance of equivocation." However, the lure of fortune, the signing of treaties between Americans and the native tribes and the advancing European settlers would eventually play a part in revealing that which Dubuque tried so hard to keep secret.

Peterson reports that even with their primitive methods of smelting lead ore in makeshift fire furnaces, the local Indians were still able to produce 400,000 pounds of lead—ten times what Pike was led to believe! Lead was so plentiful in this area that by 1815, there were twenty crude Indian smelting operations in the Galena region alone. Petersen adds that by 1816, the first flatboat carrying lead from the Fever River in Illinois made its way south via the Mississippi, sent by a certain George Davenport to the port of St. Louis. Within five years, these small vessels loaded with smelted lead ingots were a common sight on the river.

The movement of settlers and miners into the Galena, Illinois area began in 1819, and the American government made the decision to post a field representative there whose job would be to supervise this region. The presence of this agent encouraged squatters from Tennessee, Kentucky, Missouri and southern Illinois to flow into the area, traveling upriver in keelboats and over land through Fort Clark, which today is called Peoria. On January 4, 1822, the first mining lease was granted, with three additional titles issued by the end of the year. In 1823, a total of nine leases were agreed upon, even though numerous settlers established operations without them.

In 1823, this region of the Midwest was largely inhabited by Indians. In Iowa, Native American tribes controlled most of the land, while in Wisconsin and Illinois, all territory north of a line drawn due west from the southern part of Lake Michigan to the Mississippi River belonged to them. The only exceptions were small parcels held by whites in Prairie du Chien and Chicago. But the movement westward by ambitious settlers would not be denied. Law and order were supposed to have been maintained by the government agent who issued and recorded land leases.

Each lease for land allowed citizens to claim 160 acres for a period of three years. In exchange, the government was to be paid a total of one-tenth the value of the mineral mined. In 1830, that tax amount was cut in half. However, the overall system proved to be unworkable, and by an act of Congress on July 11, 1846, the land was "brought into the market and sold." From the very beginning, activity in this mining region was remarkable. For example, on July 1, 1825, one hundred miners were working at the Fever River lead mines. Just over a year later, by August 31, 1826, the total had increased to well over four hundred. These numbers would climb steadily upward for years to come.

According to Petersen, the lead-mining history in this region of the country from 1823 to 1848 can be divided into three distinct stages. The first period, from 1823 to 1829, saw a steady and rapid growth of settlers and mining productivity. From 1829 through 1835, historians note a decline in production, but the population continued to be strong. The time from 1835 to 1848 is seen today as the stage of the greatest mining activity, when annual lead production rose from eleven million pounds to fifty-five million pounds, with a total value of nearly $1,700,000.

Petersen offers additional statistics to support his conclusions. In 1824, for example, 175,220 pounds of lead were mined, and by 1829, the total rose sharply to 13,994,432 pounds. Two economic factors—the previously mentioned tariff and a saturated lead market—led to a time of depression and uncertainty in the mining industry, yet production up to March 31, 1829, was still 31,764,862 pounds, and the total amount from that date to 1834 was 34,901,205 pounds. Petersen insists that "had conditions been more favorable," given the proportions of the increase during this time, the amount of extracted lead could have reached close to 100 million pounds by the end of 1834. From 1835 onward, lead-mining production continued to increase. From 1835 to 1844, 192 million pounds were processed, and in the following four years, from 1844 to 1848, the total stood at 204 million pounds, an amount that Petersen emphasizes almost equaled the total production of the previous twenty years.

Many towns sprouted up in this lead-mining district, most along the banks of rivers. Galena, Illinois, for example, was situated on the west bank of the Fever River. Located only 500 miles from St. Louis, 200 miles from Chicago and 440 miles from Green Bay, the settlement of Galena soon became the jumping-off point for prospectors who were searching for ore to mine. Dubuque, Iowa, and Potosi and Cassville in Wisconsin successfully competed with Galena because of their favorable transportation locations along the Mississippi River. Other towns such as Dodgeville, Platteville, Shullsburg, Mineral Point and New Diggings in Wisconsin became important inland ore locations.

The first steamboat made its way up the Mississippi to the lead mines in 1823. In doing so, the vessel *Virginia* proved that river navigation was indeed practical and sparked a proposal for a general improvement of the river's channel and the elimination of hazardous rapids. Petersen reports that the then superintendent of mines, Lieutenant Martin Thomas, recommended to Secretary of War John C. Calhoun that $30,000 be spent to make the necessary changes on the Mississippi. He argued that not only would these improvements benefit the commercial transportation of lead, but they would also make it easier for the government to move troops and supplies. Following in the wake of the *Virginia*, the steamboat *Rambler* made a "hasty trip to St. Peter's" and returned to St. Louis in September. The presence of the *Virginia* and the *Rambler* marked the beginning of the steamboat era on the upper Mississippi River.

The following three years saw limited lead ore production; most of the ingots were transported downriver in keelboats. Petersen reveals that in 1825, only three steamboats traveled up to St. Peter's: the *Rufus Putnam*, the

Lawrence and the *Gen. Neville*. The *Gen. Neville* returned to St. Louis with a primary cargo of pelts and furs for the Columbia Fur Company. In 1826, steamboats *Lawrence*, *Mexico*, *Scioto* and *Eclipse* visited the upper Mississippi, but Petersen does not say whether they transported any lead on their return trips south, except to write that "prior to the year 1827, scarcely a dozen different steamboats frequented the waters of the upper Mississippi and not one of the number was engaged solely in the lead trade." However, within a year, activity at the Fever River lead mines reached a "fever" pitch, and the history of the area soon would be changed forever.

Petersen reports that in the spring of 1827, "a wild stampede for the Fever River lead mines began," presumably because of a new ore strike. The impact of this increased activity was felt as far south as Missouri, where "traffic at the St. Louis levee was immense." Petersen provides greater details in the following description of the scene:

> *By March 15,* [1827,] *one steamboat had already left for the mines, three were advertised, and it was estimated that at the rate people were pouring into the city, several thousand would head northward before navigation ceased. By the close of the week ending April 19, the wharf was constantly thronged with restless knots of determined men, anxious to reach the mines and try their fortune. Eyes constantly strained to catch the first glimpse and ears listened for the hoarse blast of an approaching boat.*

But that wasn't all. The Fever River mines were yielding huge quantities of lead, and the resulting frenzy spread rapidly. Peterson continues:

> *The* Indiana *and the* Shamrock *came into port on April 19 heavily freighted with lead, and several keelboats arrived the same day with similar cargoes. Before the close of navigation, almost 7,000,000 pounds of lead had been mined and transported down the river. During the season, the steamboats* Lexington, General Hamilton, Indiana, Lawrence, Mexico, Muskingum, Mechanic, Pilot, Shamrock, Scioto, Velocipede *and the* St. Louis and Galena Packet, *had visited the upper Mississippi, and practically all of them had plied fairly regularly in the lead trade. This was a larger number than had visited the upper Mississippi in the previous four years.*

Petersen also quotes an excerpt published in the Niles (Illinois) *Weekly Register* newspaper to describe what was happening in Galena, a town he characterizes as already "boasting a weekly paper and seven hundred aggressive pioneers" as a result of this increased lead production:

> *There are forty-two stores and warehouses; twenty-two porter cellars and groceries; a goodly number of lawyers and physicians, and a general assortment of mechanics, &c. The number of dwelling houses and stores is 195; and 46 new buildings are going up. There have been 75 steamboat and 38 keelboat arrivals since 1st March. Almost* eighty million *pounds of lead ore were exported during the year ending 1st of June last. The population in the neighborhood of the mines is estimated at 10,000* [emphasis in original].

Petersen adds that goods from the northbound steamboats were constantly being dumped onto Galena's wharves, and replacement cargoes of lead were just as quickly loaded for the returning southbound journey to St. Louis, "which would sink the boat to her guards." In particular, five steamboats brimming with lead left Galena in one day in March. At the end of the navigation season, lead-bearing steamboats had made over one hundred trips.

Nearly all the lead that was mined and shipped to eastern markets from 1823 through 1848 traveled down the Mississippi River. Despite the frequently high freight charges, the natural obstructions the river had to offer and the short navigation season, the "Father of Waters" provided the most logical route for shipping the ever-increasing output of lead to market. According to Petersen, the impact of mining in the region of the upper Mississippi River on the development of the steamboat industry had four components:

> *(1) Thousands of hardy immigrants were encouraged to venture into lands that, in some cases, were located far beyond the existing frontier, causing it to be extended. (2) This increased population brought with it the demand for imported supplies and materials needed in a region that was not yet self-sufficient. (3) The demand for lead created an opportunity for steamboat owners and operators to make a profit by bringing the smelted product to market. (4) Above and beyond the transportation of lead, the large number of steamboats that navigated the Mississippi also ferried the steady waves of immigrants who were seeking their fortunes in the "Northwest," and with them the supplies they needed and the goods they produced for down river markets.*

From 1830 to 1834, the previously mentioned tariff kept the production of lead down, but still, the average number of steamboat voyages from the area was eleven. The Black Hawk War in 1832 also had an impact on river traffic, as only six steamboats dared to enter the war zone. After Black Hawk was defeated and the territory in Iowa that his tribe controlled was ceded to the United States, the next phase of lead ore mining and production would soon begin.

Concluding his research, William Petersen offers this observation about the role the steamboats played in the early development of the Midwest, based in large part on the mining of lead:

> *Had the steamboat failed to establish itself as the dominant factor in transportation and in communication with the upper Mississippi as early as it did, the construction of railroads might have taken place a decade earlier and a picturesque phase of upper Mississippi Valley life would have been lost to posterity. After 1848, the steamboat was so entrenched in the economic life of the region that it was able to wage a thrilling, albeit losing, battle against the railroad.*

Even before the Black Hawk War of 1832, impatient pioneers set their sights on the possible wealth that lay beyond the western banks of the Mississippi River. After the treaty with the well-known chief was signed and ratified, a new era in American history began as enterprising miners staked their claims and processed lead ore. Among them were individuals who eventually would be associated with the development of Dubuque.

Chapter 8
White Men Ante Portas

By Tawni Bauer

Editor's note: The following research is adapted from Antique Dubuque: 1673–1833, *written by Rev. M.M. Hoffman and published in 1930. The author was a member of a pioneer Dubuque family and taught at Loras College for twenty-six years.*

Around 260 BC, Hannibal, the famous Carthaginian general, and his armor-dressed elephants were on their way to the Alps. He was on a journey to reach the Roman Empire that he hated and wished to eliminate. "Hannibal ante portas," meaning "Hannibal is at the gates," was the phrase spoken by the startled Romans when they saw Hannibal on the outskirts of the city after he crossed the Alps in the dead of winter. Some two thousand years later, the Meskwaki Indians perhaps felt a similar fear, or "white men ante portas," when the pioneers were just across the Mississippi River, waiting to invade their land. The Indian chiefs knew that they couldn't keep the settlers out much longer. The whites, waiting to get in, expected that the Indians would eventually be removed.

As early as 1819, Colonel James Johnson of Kentucky was illegally mining and trading lead from the Fever River area in northwest Illinois. Finally, in 1822, he bought a lease from the American government. He employed some white men and black slaves. He also started using a supply of tools and operated keelboats to haul the mined mineral. Many other white men hurried to the Fever River area. By August 31, 1826, 453 leases had been

James Langworthy (1800–1865). Head of a successful mining family in Dubuque, his knowledge of local terrain aided the American army during the Black Hawk War. Langworthy later owned and operated the Shot Tower (see Chapter 20). *Dubuque County Historical Society.*

issued, compared to only 13 in 1823. The total number of leases would dramatically increase into the thousands in the coming years.

In 1829, James L. Langworthy was granted permission by the Meskwaki Indians to explore the Dubuque area. Some twenty-five years later, Langworthy and his three brothers, Edward, Lucius and Solon, would become important members of the Dubuque Harbor Improvement Company. This business filled in the swampy land on the western bank of the Mississippi River around where the Shot Tower would one day be built to encourage economic development in that area. In 1830, Lucius Langworthy observed, "There was not then a white settler north of the Des Moines [River], and west of the Mississippi to Astoria, on the Columbia River, with the exception of the Indian traders." There weren't many white settlers in the area around Dubuque, either, but when they were finally allowed into the mining area, they came like a flood.

In 1830, the Meskwaki Indians left their lands around Dubuque and migrated south because they feared that the Sioux and Menominee tribes would attack them. These groups had killed many of the Meskwaki warriors in the past. This was when the whites made their first attempt to claim the Dubuque lead area for their own. As soon as they heard about the Meskwaki fleeing the area, one hundred men from the Galena district in Illinois crossed the Mississippi River and went to work in the Dubuque lead mines.

The men who traveled to the Dubuque area came up with a set of laws to avoid conflict between miners as they began to take lead from the rich deposits. The laws were completed on June 17, 1830. They are as follows:

> *We, a committee having been chosen to draft certain rules and regulations, by which we, as miners will be governed; and, having duly considered the subject, do unanimously agree that we will be governed by the regulations on the east side of the Mississippi River, with the following exceptions to wit: Article I—that each and every man shall hold two hundred yards square of ground by working said ground…Article II—we further agree that there shall be chosen by the majority of the miners present, a person who shall hold this article, and who shall grant letters of arbitration, on application being made, and that said letter* [of] *arbitration shall be obligatory on the parties concerned so applying. To the above, we the undersigned subscribe J.L. Langworthy, H.F. Lander, James McPheeters, Samuel H. Scoles, E.M. Urn.*

Unfortunately for these miners, the land of Iowa still belonged to Black Hawk and his tribe in 1830, and there was an agreement between the U.S. government and Chief Black Hawk that no whites would cross the Mississippi River, the natural boundary that separated the two races. Even though this agreement was in place, miners who wanted to seek their fortunes just couldn't stay where they belonged.

In July 1830, Colonel Zachary Taylor, who was posted at Fort Crawford, arrived at the Dubuque mines. He instructed the miners, including the Langworthys, to cross back over the Mississippi River into Illinois. The settlers were told to leave within a week or suffer the consequences. All but four of the white miners left, and they were taken prisoner. Soon after the whites had been moved off, the Meskwaki people returned to the mines and sold the lead that had been mined while they were away. This made the white miners very angry.

Black Hawk (1767–1838). The noted Sauk war chief's Indian name, Makataimeshekiakiak, is interpreted in English to mean "be a large black hawk." *Dubuque County Historical Society.*

The Meskwaki Indians had been killed by the Menominee in battles in 1831. The Meskwaki then left the Dubuque mining area once more for fear of being attacked again. Soon after, a few white explorers ventured back across the river to the mine area once the Meskwaki were gone. The U.S. government had not purchased or acquired the mine lands; therefore, the War Department sent in troops under the command of Lieutenant Jefferson Davis (later president of the Confederate States of America) to clear the area of white settlers once again. Lieutenant Davis and his troops, along with Lieutenant J.R.B. Gardenier, set up their camps in the mining area to keep the white settlers from reentering. They stayed there until 1832, when the Black Hawk War broke out. Then they were called up to join other military units in the battle against the Indians.

Lieutenant Davis wrote about his time in Dubuque County. He said that on occasion, he traveled into the country to watch the movements of the Indians. When he was asked how far he had traveled, he stated that sometimes he went as far as the Maquoketa River and back to the Dubuque area in the same day. Considering the rough terrain of that time, Davis's trips were seen as great accomplishments.

With the soldiers gone to fight the Black Hawk War, the land was free of any kind of military presence, and the white settlers who were watching across the Mississippi River knew this. Once again, some of the same men returned to the Dubuque mine area, including James, Lucius and Edward Langworthy. Between two and three hundred people settled in the area to take advantage of the fact that no one was there to keep them out.

In December 1832, a man by the name of Marmaduke S. Davenport was given orders by the War Department to remove these recent intruders. Most of them left right away, but because of the huge number of miners, they were given ten days to leave. The miners were very disappointed and didn't want to give up their claimed land again. They prepared and signed a petition that would be given to the secretary of war. They explained that they had built cabins and other buildings to shelter them from the cold winter. They also made sure the U.S. government knew that they were willing to pay rent for the land that they settled on. They said this in their petition:

> *It is with regret we learn that the Government issued its orders to have us expelled east of the Mississippi. We have made some improvement, such as built us cabins &c., to shelter us from the inclemency of the winter, peculiar to this latitude. We have not come here as intruders. It is not our intention*

Edward Langworthy (1808–1893). Mined lead with brother James, participated in Iowa's constitutional convention, served as a county commissioner and alderman. After the Panic of 1857, he became the director of the First National Bank of Dubuque. *Dubuque County Historical Society.*

> *to wrong the Government, for the mineral or ore that is now raising at these mines will not be manufactured into lead before next spring. And then we will hold ourselves in readiness to pay such rent as is, or may be, established by law from time to time, on the upper Mississippi lead mines.*

The miners said that if they were kicked out of the area, they would then become homeless. They also mentioned that they would have no money because they had not yet gotten paid for serving in the Black Hawk War. The miners made sure they mentioned that they had not seen any Indians around the area, either:

> *We would also state that if the order of the department be promptly executed, it would leave us, (and some with large and helpless families) in a suffering condition, houseless and penniless, in the dead of winter; penniless we say, not as yet having received our pay after a long and aggravated summer's war. And as for danger being apprehended* [feared] *by Government of a collision between whites and Indians, we would say such apprehension is groundless, for there is not, nor has there been any Indian within one hundred and fifty miles of this place, to our knowledge since the treaty.*

Unfortunately for the miners, there was no pity felt for them by the U.S. government due to the fact that it knew the land was not theirs to settle until the United States Senate ratified the treaty with Black Hawk. Lieutenant George Wilson was then sent from Fort Crawford, Wisconsin, to make the whites leave and to prevent any trespassers, mainly whites, from entering the Dubuque area. Once Wilson completed his assignment, the land was again free of the white man.

On February 22, 1833, Marmaduke Davenport, who was the Indian agent at Rock Island, received word that eighty to one hundred people had entered the Dubuque area and were back mining and smelting lead ore. Some had built homes in hopes of keeping their land titles until it would become completely legal for them to live there. Unfortunately, these illegal miners found themselves helpless and out of luck one more time because they were all sent back across the Mississippi River. However, Lieutenant Wilson was not as strict with the task of keeping all the white men out of the Dubuque mining area.

It was said that Wilson allowed a handful of them into the mines because he wasn't serious about his job. He became disagreeable toward his task and received a three-month leave of absence. Before he left, Lieutenant Jefferson Davis and Lieutenant J.J. Abercrombie and their troops traveled from Prairie du Chien to take over for Wilson and carried out the expulsions. Many of those miners who were kicked out never returned to the area, although a number of others moved onto the islands in the Mississippi River that happened to be right across from the mines. They had a strong belief that the mineral they had extracted from the mines was rightfully theirs.

Lieutenant E.F. Covington was also stationed in the Dubuque mine area. The miners really hated this man because he was very firm about keeping them out. Although he was strict, the miners continued their defiance toward the rules to stay on the east side of the Mississippi. It seems that the lure of mining lead was so strong that it would take an army to keep the miners out of the area.

When Chief Black Hawk attacked white settlers on the east side of the Mississippi, the American government sent in troops and defeated him. Soon after, a treaty was signed, and Black Hawk gave up the lands he controlled on the western shore of the Father of Waters—the Mississippi River. After the treaty was ratified by the U.S. Senate in June 1833, the mines in Dubuque and the land beyond were finally and legally opened to the settlers. White men ante portas.

Chapter 9

Coming to Iowa: The Lucius Langworthy Lectures

By John Adelmann

Editor's note: The following narrative comes from two lectures given by Lucius Langworthy, one of the early white settlers who traveled to eastern Iowa in the late 1820s seeking his fortune in the lead-mining business. By the 1850s, Langworthy and his brothers were major figures in the development of Dubuque's downtown business area. They created the Dubuque Harbor Improvement Company, a business from which George W. Rogers secured a mortgage in 1856 to construct his Shot Tower by the Mississippi River. Langworthy was invited to share his remembrances for posterity with members of the Dubuque Literary and Scientific Institute on December 18, 1854, and on February 26, 1855. His remarks were recorded in the Iowa Journal of History and Politics *8. Spelling and punctuation appear as in the original.*

In his letter of acceptance to the Committee of the Dubuque Literary and Scientific Institute, Lucius Langworthy cautioned that his recollections of events that took place some thirty years before might be somewhat incomplete but may still be of some value to future historians:

> *Permit me to say that they contain only the prominent incidents of a frontier settlement, consequently in the events narrated much of the minuteness and connectedness of the framework of history will be wanting. I am aware that*

they only possess the merit of originality, so far as they contain portions of the unwritten history of the times.

In his opening remarks before the committee, Langworthy again expressed his fear of not being able to give an accurate account of his experiences and asked for the members' understanding. He cautioned:

I hesitated some time before accepting the kind invitation to address you upon this occasion. I have to call up facts of our past history from a somewhat treacherous memory, and prepare a sketch without a written data, stirring incidents, or interesting details, original in all its parts, and truthful in all its relations. But relying on your liveliest charities, I thus undertake the task.

He then began his lecture with a general description of the land he saw upon his arrival in this area in 1827:

There were no white inhabitants settled upon all the region north and west of the Illinois River. Thirteen miles square of mining land had been ceded to the government of the United States, by the Winnebago Indians, at Galena; a few trading posts were established along the Mississippi, at various points above St. Louis, forts were erected and garrisoned at Warsaw, then Ft. Edwards, Ft. Armstrong, at Rock Island, and Ft. Crawford, at Prairie du Chien. Indian Villages lined the banks of the ocean [sic] *stream, at different places, and mere trails united us with St. Louis and Central Illinois.*

Langworthy described for his audience the native peoples who were already living in what he characterized as "uncharted and uncivilized" land:

The Indians were immensely numerous, jealous of the white people who were approaching their borders, and ready upon any provocation to fly into open hostilities against them. Many renowned chiefs, such as Black Hawk, and Keokuk, stood at the head of the respective bands of the different nations to inspire dread and apprehension among the mingling masses, as the first wave of civilization rolled onward to the western wilds.

He also explained the difficulties that early settlers faced in securing supplies and traveling to reach the lead mines:

> *At that time no steamboats navigated the Mississippi, except, occasionally one, laden with Government stores for the different garrisons along the river. Supplies for the new settlers had to be transported from Ohio to Kentucky, in keelboats, and a journey to the lead mines then, was like a journey now over the mountains to Oregon, and California.*

Despite the potential for success, Langworthy initially thought his mining expedition into the unknown would be temporary at best. He said, "No thought was then entertained that this mining region would ever become the home of permanent settlements, or useful for purposes of agriculture. It was deemed a sterile frozen region." He also acknowledged the significance of the Galena settlement in the region's development some thirty years before Dubuque was known as a small mining community, let alone a town:

> *Galena, had, however, a place on the maps, and became of considerable commercial importance long before Dubuque was known, except by the French and Spanish traders, who, for barter with the Indians for their furs and peltry, had long before penetrated the wilds of North America. But even they had gone onward to the better hunting grounds in the great plains and mountain passes of the west.*

In the fall of 1824, Langworthy's brother James arrived in Galena from St. Louis, and together with other settlers, they opened the mines of Hardscrabble, located about nine miles north of Galena. Ore was soon struck, and with it came arguments and fights. Langworthy explained how the town got its name: "Hardscrabble diggings…received the name from a terrible encounter which took place between contending parties for possession of a lode, in which hard blows, rifles, and Bowie knives, were freely used."

Lucius Langworthy joined his brothers three years later, along with another brother and two sisters who traveled from Quincy, Illinois. Their trip took them a month. Traveling in a boat called a pirogue, Langworthy says he purposely mentioned the time it took to complete the journey in order to emphasize its challenges:

> *We embarked at Quincy, Illinois, in a pirogue, and were thirty days on the voyage. A pirogue is a kind of intermediate craft, between a canoe and a keelboat. The name is French, and signifies the kind of boats used by the early voyageurs to transport their furs and effects over the shoal waters and*

> *rapid streams of the western wilderness. I mention the time occupied in our journey hither, in order to show some of the difficulties of settling this new country at that early period. Think of a boat's crew, with several ladies on board all unaccustomed to the river, being compelled to work a boat with poles, and oars, against the swollen current of this mighty stream, in the hot weather of June, sleeping on sand bars, or anchored out in the river at night to avoid the musketoes* [sic], *or lurking Indians, living upon salt pork and dry biscuit, coffee without cream or sugar…making only about eight miles coverage per day.*

Despite these hardships, Langworthy evaluated the endeavor's potential, saying, "But this was then the land of promise, as California has been since then."

By 1830, intertribal hostilities broke out between the Sauk and Fox and a war party of Sioux that proved to have an impact on the future of lead mining in this area. A dozen Sauk and Fox chiefs and their respective groups were traveling to attend a treaty conference in Prairie du Chien that Langworthy said had been called by "United States Commissioners." But at Cassville Island, the Sioux attacked, leaving all but two Sauk and Fox Indians dead. Langworthy observed, "The tribe, now in great confusion and alarm, left the place and the graves of their Fathers, mostly never to return, and thus these mines and this beautiful country was left vacant and open for settlement; for previously the Indians would allow no one to intrude upon their lands."

Some thirty years earlier, the French explorer Julien Dubuque had worked the lead mines in the area now called Dubuque, Iowa. Here, Langworthy provided a brief character sketch of the famous Frenchman, offered reasons why he lived such a relatively short life and explained why some Indians saw him as a possessor of "powerful medicine":

> *There were mines of lead worked here as early as 1800, by the natives aided by Julien Dubuque, an Indian trader…*[Dubuque] *adopted their habits and customs, married into their tribe, and became a great chief among them. He is said to have been of French and Spanish parentage, of small stature, greatly addicted to the vices incident upon the co-mingling of Spanish and Indian races in America, and a great medicine man. He would take live snakes of the most venomous kinds into his arms and bosom, and was consequently regarded by the Indians with a superstitious veneration.*

The Langworthy party crossed the Mississippi River in June 1830 and prepared to stake their claim on the lead-rich land on the western shore. This pivotal moment in local, Iowa and even national history, made possible because the native population had abandoned the area, was described this way:

> *We crossed over the Mississippi at this time, swimming our horses by the side of the canoe. It was the first flow, or the first tide of civilization on this western shore. There was not a white settler north of the Des Moines* [River], *and west of the Mississippi to Astoria, to the Columbia River, with the exception of Indian traders. The Indians had all along guarded this mining district with scrupulous care. They would not allow the white people to visit the place, even to look at the old grass grown diggings of Dubuque which were known to exist here, much less would they permit mining to be done, or settlements to be made. It was like the ancient Canaan, to the Israelitish leader, "a forbidden ground."*

Now able to explore the land once denied to white settlers, Langworthy visited the grave site of Julien Dubuque, who had died twenty years earlier, and described his burial plot:

> *A stone house surmounted by a red cedar cross, with a leaden door, was placed around his grave, which may still be seen, though in a dilapidated condition. When I first visited his tomb in 1830, the remains of two Indian chiefs were deposited within, I suppose as a mark of peculiar distinction. On the cross is inscribed the following, in French, which being translated*

Artist C.L. Trudell's 1907 rendering of Julien Dubuque's first grave site erected by his Indian mining partners following his death in 1810, as described by Lucius Langworthy after his visit to the tomb in 1830. *Dubuque County Historical Society.*

> *literally is: "Julien Dubuque, miner, of the Mines of Spain, DIED* [sic], *this 24th day of March, 1810, aged 45 years and 6 months."*

The territory Langworthy and his party were walking on had been hastily deserted by the Indians, yet their presence was still being felt: "The country had just been abandoned by the RED MEN [*sic*], their moccasin tracks were yet fresh in the prairie trails along which the retiring race had fled on their mysterious mission westward, and the decaying embers were yet cooling on their deserted hearths within their lonely and silent wigwams."

Langworthy then made reference to the ground nearby where the settlement of Dubuque later would be established and described the site along the Catfish Creek shoreline where natives had once lived and worked, mining and smelting lead ore with their partner Julien Dubuque:

> *Where Dubuque now stands, corn fields stretched along the bluffs, up the ravines, and the Coule valley, and a thousand acres of level land skirting the shore, was covered with tall grass, as a field of waving grain. But the stalks of the corn were of the last year's growth, the ears had been plucked, and they withered and blighted, left standing alone MOURNFUL REPRESENTATIVES of the VANISHED RACE* [emphasis in original].

He continued his lecture with a sober description of the forsaken Catfish Creek compound that revealed the extent to which the Meskwaki had developed a permanent settlement:

> *A large village was then standing at the mouth of Catfish Creek, silent, solitary, deserted—no one remained to greet us, but the mystic shadows of the past. About seventy buildings constructed with poles, and the bark of trees remained to tell of those who had so recently inhabited them. Their council house, though rude, was ample in its dimensions, and contained a great number of furnaces, in which kettles had been placed, to prepare for the feasts of peace or war. But their council fires had gone out. On the inner surface of the bark there were paintings done with considerable artistic skill, representing the buffalo, elk, bear, panther, and other animals of the chase, also their wild sports of the prairie, and even their feats in wars where chief meets chief, and warriors mix in bloody fray.*

A sense of loss is evident in Langworthy's listing of the artifacts he recognized as links to the past. He wished there could have been some way to safeguard them. Unfortunately for history, the site was soon destroyed: "Could the place have been preserved on the canvas, or by the Daguerrean art [period photography], it would have been an interesting relic, but nothing now remains of it, and but a few know that such a place ever existed. It was burned down in the summer of 1830 in a spirit of vandalism, much to the regret of the new settlers."

The reader will remember that in 1830, the soil Langworthy and his fellow settlers were exploring was still considered Indian territory, and not until after the defeat of Chief Black Hawk in 1832 would the American government allow whites to claim it and work it. In fact, the U.S. Army detachments were dispatched to several settlements to remove white squatters and protect Indian rights. However, ambitious miners, including Langworthy and his family, would not be denied; the results of their labors proved to be very profitable:

> *Soon after this* [crossing the Mississippi into Iowa territory], *a number of miners crossed over the river and possessed themselves of these lands…their mining operations proved eminently successful. Many discoveries were made, which bid fair to realize the most romantic ideas of the wealth previously thought to be stored within the subterranean vaults of these mines. It was known that Dubuque had obtained immense quantities from them, and the accounts which the Indians gave of them, served to strengthen the belief that the veins of ore here, were inexhaustible. The first lode discovered was of great value, and confirmed this opinion.*

On Independence Day 1830, Captain Zachary Taylor, then commander of a military detachment in Prairie du Chien, ordered the miners off the land, claiming that it had not yet been purchased from the Indians and, as such, was under the jurisdiction of the War Department. According to Taylor, treaty stipulations demanded the miners vacate immediately. Langworthy remembered the miners' reply to Taylor and his comeback to them, saying:

> *They declined doing this, telling the Captain that* he *must surrender this time. They urged that they had occupied a vacant country, had struck some valuable lodes, that the land would soon be purchased, and that they intended to maintain possession; to which, Zachary Taylor replied, "We shall see to that my boys"* [emphasis added].

Zachary Taylor (1784–1850). Commander of the military detachment from Prairie du Chien who evicted encroaching miners from Indian territory before the Black Hawk War. "Old Rough and Ready" later became the twelfth American president from 1849 to 1850. *Dubuque County Historical Society.*

True to his word, Taylor, whose nickname was "Rough and Ready," ordered his troops, in Langworthy's words, to "make the miners at Dubuque walk Spanish." He added:

> *Anticipating their* [Taylor's and his troops'] *arrival, they* [the miners] *had taken themselves off, for at that early day they believed that "rough" would be "ready" at the appointed time. The miners were anxiously peering from the high side of the bluffs on the east side of the river as a steamer came in sight bringing the soldiers, who were landed on the west shore. The military force was stationed permanently at Dubuque, and the Indians, venturing back to the place sure of safety and protection against their inveterate enemy, the Sioux and other intruders, were encouraged to mine upon the lodes and prospects which the white people had discovered.*

An anguished Langworthy then related how the Indians took advantage of the settlers' hard work and what the ultimate cost of that aborted venture was:

> *From one mine alone, the Indians obtained more than a million pounds of ore, in which they were assisted by the traders and settlers along the river, with provisions, implements and teams, while the discoverers, those who had opened these mines again after they were abandoned by them and the*

> *Spanish miners more than twenty years, were compelled to look across the water and see the fruits of their industry and enterprise consumed by the Indians. We lost in this manner, more than twenty thousand dollars' worth of mineral, which was taken from one lode by them.*

Interesting is Langworthy's description of how Native Americans processed lead in the Dubuque area. The Indians, he reported simultaneously, "do not mine or labor in any way" and also mined "on special occasions":

> *When the Indians mined, however, which was on special occasions, there were often fifty or a hundred boys and squaws at work, on one vein. They would dig down a square hole, covering the entire width of the mine leaving one side not perpendicular, but at an angle of about forty-five degrees, then with deer skin sacks attached to a bark rope, they would haul out along the inclining side of the shaft, the rock and ore. Their mode of smelting was by digging into a bank slightly, then put up flat rocks in a funnel shape, and place the ore within, mixed with wood; this all burned together, and the lead would trickle down into small excavations in the earth, of any shape they desired, and slowly cool and become fit for exportation.*

Other research reveals that this primitive method of smelting lead extracted about half the available mineral from the total. Initial heat from Indian fires melted the lead, but as their temperatures subsided, the metal cooled and turned back into a solid, or became a worthless mixture of lead and ash in the dying embers. White miners called this byproduct "clinkers." Some "clinker-like" material was later re-smelted in hotter furnaces, and more of the remaining lead was separated from the ash. Despite their limited smelting ability, however, the Indians still sold and traded processed lead ingots in large quantities, a clear indicator that the ore was plentiful.

In a reversal of roles just two years later, the same troops who evicted squatters from Indian lands were now being called upon to fight Black Hawk and his coalition of tribes who were attacking white settlers in a vain attempt to regain land already ceded to the United States. Langworthy remembered this ironic turn of events: "Zachary Taylor, with his forces and the volunteers under Col. Dodge, composed in part of those miners thus despoiled of the fruits of their first labors at Dubuque, were among those who followed the Indians and finally defeated them at Wisconsin and at the Bad Axe, in the fall [August] of the same year, putting an end to that war."

With a dash of smugness tempered by time, Langworthy then told how, after years had passed, the Dubuque miners were able to put Colonel Taylor's actions against them in perspective and added that a more mature, now-president Taylor reconsidered the behavior of the once-pesky white squatters when he and Langworthy chatted at the White House in Washington:

> *They* [the Dubuque miners] *recollect Col. Taylor well, indeed almost too well, though only as a faithful and brave officer. What he did at Dubuque, although it robbed them of the rewards of their labor, it is supposed he thought to be his duty under his general orders, and as for his conduct in the wars of our country, since that period, that is a portion of American History. Long after this, while conversing with him in the White House, the old veteran alluded to these early scenes of his life with great gaiety, and declared that those miners at Dubuque were worse to manage than the Seminoles or even the Mexicans.*

In September 1832, a treaty was signed at Rock Island by General Winfield Scott and others representing the United States, and the Black Hawk purchase was made official. Langworthy said the area "included all the country bordering on the west side of the Mississippi River, comprising the eastern portion of our State. About this time those who felt an interest in the mines of Dubuque, returned to take possession of their former discoveries." However, even though the treaty was agreed to and signed, white settlers who returned to the area were once again forced to vacate their claims because of a technicality that left them stranded in the winter of 1832–33:

> *Many fine lodes and prospects were discovered and considerable lead manufactured up to about January, twenty-fifth, 1833. But in January the troops were again sent down from Prairie du Chien, and removed the settlers the second time, merely because the treaty by which the land was acquired, had not been ratified by the United States Senate, a formal act that everyone knew would take place at the earliest opportunity. This was a foolish policy on the part of the Government, and operated peculiarly hard on the new settlers, who were thus obliged to leave their cabins in the cold winter of 1832–33, and their business also until spring.*

The settlers had little choice but to follow orders, and some remained huddled on a river island with their ore until spring:

> *Many re-crossed the river and did not return. We repaired to the Island, and there erected temporary buildings to await the ratification of the treaty. Having about three hundred thousand pounds of lead on hand, and being uncertain what would be the orders of the military regarding this kind of property taken from land not yet fully owned by the Government, we removed it also to our island home, and remained by it until spring, the soldiers meantime occupying our warm and comfortable dwellings at the mines.*

The spring of 1833 saw miners who had endured the winter return to the ore fields. The treaty now official, miners could stake legal claims, but with one development that no one appreciated—a production tax:

> *In June, 1833, Mr. John P. Sheldon, arrived with a commission from the department at Washington, as Superintendent of the mines, the military force having been previously withdrawn, and the treaty confirmed. He proceeded to grant written permits to miners, and licenses to smelters. These permits entitled the holder to the privilege of staking off two hundred yards square of land wherever he chose. If not occupied by others, and have peaceful possession, by delivering his mineral to a licensed smelter, while the smelter was required to give a bond to the agent conditioned to pay, for the use of the Government, a fixed percentage of all the lead he manufactured. Mr. Sheldon continued to act in this capacity only about one year, for he could not be the instrument of enforcing this unjust and unwise policy. He saw that these men, like all other pioneers who by their enterprise were opening up a new country, and fitting it for the homes of those who follow their footsteps, should be left, by a wise and judicious system, to the enjoyment of their hard earnings. The hidden wealth of this earth, its pine forests and surface productions, should alike be offered freely to all who penetrate the wilderness, and thus lay the foundation of future societies and States.*

At this point in the lecture, Langworthy offered his personal opinion about the government's revenue-generating scheme that was imposed on the settlers:

> *It has been the policy of our Government, at various times, to extract rent for all mineral, or pine lumber taken from public lands; which policy is wrong and should be forever abandoned; for the early settlers have privations and hardships enough without encountering the opposition of their own government, especially these miners, many of whom had labored for years on*

the frontier, cut off from the enjoyments of home and all the endearments of domestic life. Your speaker was one of these, being thrown in early life upon the crest of the wave of western emigration, often beyond the furthest bounds of civilization, and not infrequently amid the tragic scenes of border strife.

Having spoken his piece, Langworthy then acknowledged the presence of a Higher Power Who had protected him and his family all these years and Who allowed him to witness the literal expansion of the American nation. However, Langworthy still chafed at the idea of that dreaded tax. Speaking of himself in the third person, he mused:

But among all such toils and dangers, a kind Providence has thus far safely guided his wandering footsteps. For thirty-seven years he has witnessed the expanding energies of the West in various stages of progress and development. In that time five new States have grown up from a homeless wilderness, and become great and prosperous, teeming with a rapidly augmenting population. Twenty-three years he has labored mostly in the mines, in different capacities, and during about half that period has toiled in the deep narrow caves and crevices, in the cold damp ground, working upon his knees, sometimes in the water, and living like many other miners, in "Bachelor's Hall," cooking his own food, and feeling secluded from society and far from the circle and associations of youthful friendship. Under such privations he felt the demand of a heavy tax by the government to be oppressive indeed, and he would be wanting in consistency and spirit if he had not on all proper occasions protested against a system that seems much more regal than republican, and which degrades the Western Pioneer to the condition of a tenant at will of the General Government.

In a final shot across the government's bow, Langworthy summarized the pioneers' contributions to the country, which he believed were valuable to the nation as a whole:

The history of the West shows that in thirty-seven years the increase of population in the five States alluded to exceeds the number of people in the whole thirteen colonies at the time of the revolution; and the West has thus added such additional strength to our government as to pay a hundred-fold for the few favors it has extended to us.

Returning to his recollection of Dubuque history and its mining industry, Langworthy then reflected on the rough-and-tumble life in this settlement by the river in its early days:

> *In 1834, the town of Dubuque continued to improve. It now first received its name by a public meeting held for that purpose, and began to assume the appearance of a prosperous business place. Many stores were erected, emigrants continued to flock hither, the mines increased in richness, and as if by magic the country was transformed from a lonely wilderness into a prosperous community.*

In several lively paragraphs, he then discussed the social mores of this emerging mining community and happily reported that with the establishment of a permanent court system, the rowdy town of Dubuque began to take on a more wholesome character:

> *At this time there were but very few men in the whole country who did not indulge in drinking and gambling. "Poker" and "Brag" were games of common pastime, while betting often ran up to hundreds of dollars at a single sitting. It pervaded all classes; the merchants and other passengers to and from St.*

Lucius Langworthy (1807–1865). The first person to use the word "Iowa" to describe the unsettled land west of the Mississippi River, Langworthy also preserved much of Dubuque's and Iowa's early history by writing and lecturing. *Dubuque County Historical Society*.

Louis while on the steamboats occupied their time chiefly this way, and it was considered no disgrace to gamble. Balls and parties were also common and it was not an infrequent occurrence for one to treat his partner in the dance at the bar, if he did not, he generally performed that delicate and flattering attention to himself. The Sabbath was regarded as a holiday and vice and immorality was prevalent in every form. Yet amidst all this there were occasional gleams of moral sunshine breaking through the clouds of dissipation, and a brighter future lay before us. In fact drunkenness and rioting had already considerably diminished, and a few years afterwards we had reason to rejoice that order had taken the place of discord. Upon the establishing of courts here, first under the jurisdiction of Michigan, then under that of Wisconsin Territory, matters assumed a more peaceful and quiet aspect.

Other challenges facing Dubuque included the issues of municipal surveying and development and how private ownership of land could be protected and encouraged. Langworthy revealed the dangers and frustrations that can surface when a clear plan is lacking:

No survey of public lands had yet been made, and in the transition from the old, to the new state of things, misunderstandings naturally arose. Under the Government rules, and regulations for the control of the mines, it was necessary to work and have mining tools almost continuously on the land claimed in order to secure possession; under the new order of things there were no uniform customs prevailing regarding the possession of property; each man formed his own standard and was governed by his own opinions. It is not surprising then, that difficulties should arise. He who has passed through all the scenes and trials incident to the settlement of a new country will not readily seek another distant frontier as a home.

Later in his talk, Langworthy discussed mining. Comparing Iowa's lead deposits to those in other tri-state locations, he offered a glimpse of what a fortunate miner could experience when he finally hit pay dirt:

The mines of Dubuque differ very much from those in Wisconsin and Illinois. There, they lie near the out-crop of the limestone strata, and the mineral is found often at a slight depth without much rock excavation; but the veins are generally small, broken and thread-like. Here, on the contrary, the dip being to the south-west, we have to penetrate a hard cap-rock of considerable thickness, varying according to the elevation of the ground. Under this cap-rock material the mineral is found in large caverns, either attached to the rocks or imbedded in the clay and ocher which fills the crevices below the natural spaces of the caves. The Langworthy lode exhibited the most astonishing specimens of lead ore ever found, perhaps, in any country. As work-men, with the owners, penetrated into this cave for the first time, a hollow sound hissed from it and the air came freshly from the west, in the direction of the vein. They began to examine the cave for material, each man carrying a lamp or candle. Passing along through various windings and narrow spaces they suddenly came to one of the subterranean vaults which was completely filled with the shining ore, lighted up and sparkling like diamonds, or lying in great masses or adhering to the sides and roof of the cave in huge cubes. This mine, in its full extent through the Langworthy, Thomas Kelly, and Cardiff grounds, has produced some ten million pounds of material.

Mining lead was a labor-intensive endeavor, but when the right area was explored and the proper depth was reached, miners could be handsomely rewarded:

When the miner sinks a shaft down into one of those mineral caves, he will generally find its course and direction to be with the cardinal points of the compass. The veins of ore, here at Dubuque run, from east to west, though these are often crossed by north and south crevices and sometimes by fissures in other directions through the rocks. Although there may be no mineral within the cave first found, the practical miner will excavate along one of these cross crevices which are generally about twelve inches wide, by blasting off the sides of the rock to a sufficient capacity to admit a small car to run along through it upon wooden rails with mineral tubs upon it. Thus he will drift or excavate north or south with a view of crossing a vein of lead as it runs from east to west. When successful, he will suddenly strike into a cavern in which the ore is deposited. As the dark and hitherto unexplored cave is lighted for the first time the object of his toil stored away in these subterranean vaults; who can describe the thoughts and feelings that fill mind and breast; perhaps he has passed years in searching for the hidden treasure. Now he has the realization of his fondest hopes. "He has struck a cave," is the cry all over the mines.

As word of a miner's good fortune spread, he would become the object of much attention and respect. According to Langworthy, "Now the smelter is exceedingly obliging and attentive to him; his associates flock together to congratulate the friend so suddenly made rich, for a mine in Dubuque, is a mine indeed."

Considering that the average Civil War infantry soldier earned fifteen dollars a month in the 1860s, fantastic fortunes could be made if and when miners in Dubuque hit the jackpot. Langworthy cited the names of several compatriots and revealed what their strikes were worth:

> *The lode discovered by our friend Thomas Levin, Esq., was one of the richest and most productive of any, yielding, up to the present time about four million pounds of ore. Besides this, there are the McKenzie, Bartlett and Stewart lodes, and Karrick mines; all of immense value; together with a great many others which have ranged in value from five to twenty-five thousand dollars.*

How much lead was in Dubuque? Langworthy gave an optimistic answer but cautioned that successfully extracting it would take a lot more than just luck:

> *The lead lands will never be entirely exhausted; on the contrary when capital, science and skill are at length embarked to develop them, it will be found that not one half of their riches has yet to be ascertained. Mining may not be justly compared to the chance throw of dice, but much depends upon good judgment in the selection of ground to explore, and energy in the prosecution of the work; for it will be seen that as the mineral all lies upon one general level, and runs from east to west, that by crossing the direction of the veins one is sure, with the aid of capital and perseverance, to succeed, at least if in a good mining district.*

Evidence of how much lead was being found in Dubuque is revealed in Langworthy's description of the smelting that was going on at that time: "Some idea may be had of the mining operations in this neighborhood from the fact that there were five blast furnaces, each smelting upon average about seventy thousand pounds per week; also many log furnaces manufacturing lead to good advantage."

By 1835, the settlement of Dubuque in the territory whose name had yet to be officially determined was continuing to expand, with the mining industry now processing and exporting about fifty million pounds of lead. However, a name from the past would soon cause problems for miners and pioneer farmers who believed they already had clear title to their lands. Langworthy recounted the situation that arose when Auguste Choteau, of St. Louis, attempted to take possession of land he claimed had been duly given to him

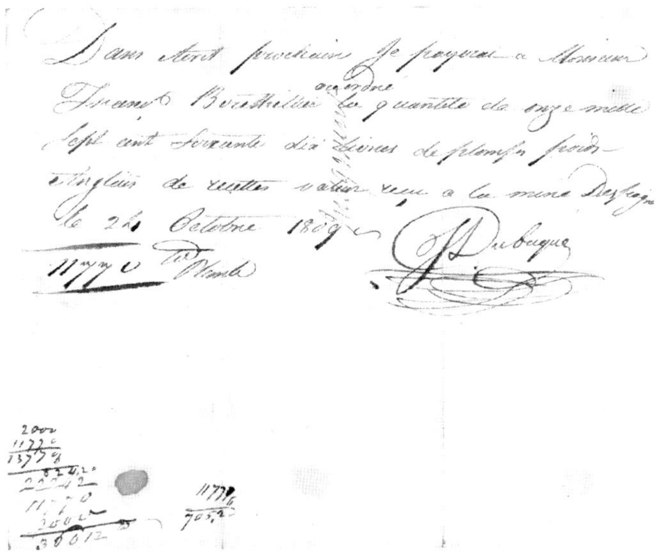

Promises, promises: "Next August I will pay…Mr. Frank Boutteillier…eleven thousand seven hundred seventy pounds of cold lead…October 24, 1809," signed by Julien Dubuque exactly five months before his death on March 24, 1810. *Kenneth W. Rendell private collection.*

by the late Julien Dubuque. The parcel of land in question included virtually the entire mining district just south of town that the Frenchman called his Mines of Spain. However, the Indians never gave Dubuque clear title to the land; all he had was the right to work it. In short, he could not sell or transfer what he never owned in the first place. This legal point did not stop Choteau, and his actions sent shock waves throughout the mining community:

> *About this time, Choteau, of St. Louis, an heir at law of Julien Dubuque, endeavored to gain possession of his claim by proposing to lease the land to miners and farmers for a mere nominal consideration; but they generally preferred the protection of the Government, and refused to acknowledge any other ownership. This claim of Dubuque, extending as it did, over the whole mining district, has been a source of much injury to the city and the country generally; for men do not like to risk their labor on lands to which there is not a perfect title. Again, the mining character of the lands served to keep them out of the market for a long time after other portions of the country had been disposed of to the inhabitants. This accounts for the newness of settlements in our vicinity, in comparison to other parts of the State.*

It took a ruling by the U.S. Supreme Court in 1853 to end Choteau's claim of ownership of the lands in question.

Du Buque Visitor.

"TRUTH OUR GUIDE, THE PUBLIC GOOD OUR AIM."

KING, Editor.] DU BUQUE, (LEAD MINES,) WISCONSIN TERRITORY, WEDNESDAY, MAY 11, 1836. [Vol. I.—No. 1.

The *Du Buque Visitor*. The first newspaper published in Dubuque and Iowa debuted on May 11, 1836. *Center for Dubuque History.*

On May 11, 1836, the first edition of the *Du Buque Visitor* newspaper was published by John King. Its motto: "Truth our guide, the public good our aim." In it was reprinted a speech given by the Honorable George W. Jones of Sinsinawa Mound. At issue was how to divide the Michigan Territory into two separate entities, with one to be called the Wisconsin Territory. While campaigning in Dubuque, Jones was asked where he stood on the issue of dividing the land. As a territorial delegate he had no vote in Congress, but he said he would support each division and would work so both areas could secure funding to help establish themselves. Jones won the election and got a government grant for the town of Dubuque to begin surveying its land. Langworthy noted, "This was an important step in the progress of the place, while the many other appropriations which Gen. Jones was enabled to procure, gave permanency to business operations."

With a bill under consideration in Congress to divide the Territory of Wisconsin, an unsigned letter printed in the *Du Buque Visitor* told of a comment made by a certain Lieutenant Lee of the board of topographical engineers, who wrote in his report to Congress in 1838 that the Iowa River was a major body of water that would one day be recognized as flowing through a great state. The anonymous writer also referred to the territory as "the future state of Iowa." This was the first time the word "Iowa" was used in connection with the territory. According to Langworthy, when William W. Corriell purchased the *Du Buque Visitor*, he changed the name of the paper to the *Iowa News*. Langworthy concluded, "Thus the name became familiar in Congress and was finally adopted for the territory west of the Mississippi."

Lucius Langworthy gives us a unique understanding of the challenges that early white settlers faced in order to gain a foothold in eastern Iowa. Their determination to mine lead kept them going, and word of their successes helped the settlement of Dubuque to grow into a thriving river town that would become the permanent home of many more settlers seeking their fortunes in mining. One day, it would also boast its own Shot Tower that would turn lead ore into finished shot for markets throughout the Midwest.

Chapter 10

My Iowa Pioneer Ancestor—Sylvester Bills Preston

By Gerda Preston Hartman

Editor's note: The author has preserved the memory of her great-great-grandfather by organizing his diaries, letters and business ledgers that have been passed down through her family. On August 1, 1986, the Dubuque Telegraph Herald *reported on Gerda's connection with her ancestors and their land in Dubuque:*

> *When Sylvester Preston brought his family to northeast Iowa in the early 1800s, this land was inhabited only by Indians and miners searching for lead. Today, the Indians and miners are gone, but the wilderness remains and so do the Prestons. Donald Preston, who now owns the farm south of Dubuque near the Mines of Spain, is Sylvester's great-grandson. When the Prestons took possession of some of the Mines of Spain, they held on to it for five generations. "We say it's been in our family since 1835," says Gerda Preston Hartman, Donald's daughter.*

My great-great-grandfather, Sylvester Bills Preston, Esq., and his wife, Anna, arrived in Wisconsin from Maryland in 1834. They were both educated in Baltimore, where he majored in chemistry, surveying and metallurgy and she in education, math and etiquette. Preston was commissioned to search for a mining site in the Platte River region near Mineral Point and then construct a copper smelter there. His children, Sylvester II, Mary Elmira, Stacy Ann (Matilda), David and Gustavus, were later sent for after their father was able to set up a home. He summed up the chances of being successful in the mining business there in a letter he wrote to his wife that was dated April 25, 1835:

> *Persons, who with but little, come here seldom fail to make money unless through their own imprudence or dissipation, although persons with considerable capital sometimes come and spend it without any favorable results. If I was so situated as to engage in it I should think prospecting for lead ore as good a chance for making money as any other in the place.*

Soon after settling in Mineral Point, Preston learned that land across the Mississippi River in Iowa was being opened for mining claims. The settlement of Dubuque was already named, and two rural townships, Peru and Mosalem, were plotted along the river in what was once Julien Dubuque's Mines of Spain. Preston took advantage of this new opportunity and settled about a mile south of Catfish Creek and Kettle Chief's mining village. His mining claims were registered with Territorial Governor George W. Jones in 1835. Preston surveyed parcels of his claim and allowed other miners to work them. If lead was discovered, he would receive a percentage. By 1836, Iowa had become part of the Wisconsin Territory.

Preston's oldest son, Sylvester II, later described the new township of Mosalem where his family moved in a historical sketch printed in the *Atlas of Dubuque County, Iowa*, and published in 1906:

> *When the Preston family moved into Mosalem Township, they found but few settlers. The country was then all open. Timber was mostly large trees which grew some distance apart. The prairie grass grew thick and high—so high that a man on horseback could not well see over it. Wild strawberries were abundant, as were the wild grape and plums. Wheat turned out well and was thrashed with flails until about 1850, when the first threshing machine was brought into the township. Wheat and some of the corn were taken to a grist mill, where the farmer waited while it was ground into flour or corn meal, giving toll for the miller's pay. Other provisions were bought in a small building where the Julien Hotel in Dubuque now stands. People devoted the greater part of their time to mining. The Indian Range and the Bonson and Brasher mines were good paying mines, but no great amount was taken from them at this time.*

Sylvester Preston was seldom home from 1848 to 1849. He traveled north to Copper Harbor, in Wisconsin on Lake Superior, to evaluate the mining possibilities there. His return trip brought him back briefly to Iowa, but then he was off again to assist the owners of the Merimack Copper Works in Missouri, just south of St. Louis, where he designed and built a smelter for the company. In his absence, his children kept him abreast of the mining activity on the Preston homestead back in Iowa. A letter from daughter Stacy Ann, written on New Year's Day 1849, updated her father on the progress of mining in Mosalem:

> *Mother told me to tell thee that she has had Charles Florgan working a month, he has been working at the "diggings" pretty much all the time as the weather has been unfavorable for doing anything else. They have been blasting through hard rock with mineral going down all the way. When they came home last night they said that they had got through into the opening on one side and got out some mineral. Sylvester II was down at the "diggings" and said that there was one chunk nearly twice as large as his head. Mother says she thinks she will not work any more at them until thee comes home.*

Sylvester Preston II maintained his inherited portion of the original family farm and mining businesses and kept a personal diary from 1869 to 1887. In reading his daily entries, I have been able to learn the meanings of words and phrases he used to describe mining tools and activities. A "gad" was a spike, any of several chisel-like or pointed bars used to pry lead ore from surrounding rock. A "mineral team" was an extra-sturdy, oak-boxed wagon pulled by horses that brought ore to the smelter. "Got pitch" describes the situation when miners dug into a mine or an area on a slant. "Clinkers" was stony matter created when leftover impurities mingled with unmelted ore and ashes after a smelting fire had gone out. "Slag" was the hardened refuse or dross separated from molten metal during smelting. A "good show" was when miners found sizable amounts of lead, as in a new vein. "Ditch around" was the practice of building up the rim of a mine shaft to prevent water and surface soil from falling into it.

Sylvester Bills Preston died on February 23, 1852, at age forty-seven. It is ironic that the man whose education enabled him to help others extract ore with smelters was absent while his family toiled away in lead mines on his own land.

Chapter 11

Historian John Klauer Adds to Shot Tower Research

"Focus on Youth" essay by Laura Mayers

While there are many articles being written about the Shot Tower by members of the Shot Tower class for the *Telegraph Herald*, I'm sure there are people asking themselves, "Where does all their information come from?" Central Alternative High School students have found a variety of sources, but possibly the most helpful have been the special guest experts and volunteer speakers who have visited our class. They have helped us to understand and appreciate our local history, and we have learned a lot of new things that we never could have found in high school history textbooks.

My classmates and I had the chance to hear about our local history from Mr. John Klauer. He discussed the significance of the Shot Tower, the abandoned lead mines located in and around Dubuque and the history of the city. Klauer worked in the Dubuque County Courthouse for almost thirty years, and he was very knowledgeable about interpreting and explaining plat maps of the city of Dubuque for us. Plats are almost like street maps. When they are compared to one another, they show changes that have been made in our city over many years. While working with these plats, Klauer also discovered interesting Dubuque history—part of which was the Dubuque Shot Tower. What I found most interesting was his discussion about the land on which the Shot Tower was constructed.

In 1856, the Shot Tower was built on a small island on the Mississippi River. Back then, all the land from the present-day Ice Harbor to Eagle Point Park was a swamp. Mr. Klauer noted that the reason why the Shot Tower was built in this watery area might have been that it was close to the river so the shot could be shipped easily and quickly. At the end of his presentation, he said he believes that our class is working on a great project. He said the public needs to know about and learn about the Shot Tower. Since it has existed for 150 years, it should be recognized for giving us a direct connection to the early days of Dubuque. No other building or structure in town can do that. The Shot Tower is a true link to the past that we can appreciate today.

Gathering information and listening to speakers aren't the only things the Central Shot Tower researchers have done. We are taking the valuable information and are creating a Shot Tower book so the public can learn what we have discovered.

Chapter 12
Shot Tower Research at the William Bowell Library

"Focus on Youth" essay by Kristi Morrison and Andy Sawvell

The National Mississippi River Museum and Aquarium is part of Dubuque's $188 million riverfront development. It allows visitors to see river wildlife and get an updated history of the Mississippi River. However, while most people stay on the main floor scoping out the water birds and the fish swimming in their huge water tanks, Central Alternative High School students can be found hard at work in the basement of the museum doing research on the Shot Tower. The library is named after Captain William S. Bowell Jr., a very successful businessman.

Bowell retired in 2001 and in 2003 donated $1 million to establish the library at the museum that now is named after him. He also set aside a sum of money that will help develop future operations. "If it's history related to Dubuque or the Mississippi River, we have it," says Denise Vondran, registrar of the Bowell Library (2005). She added, "Not only does the library have books donated by Captain Bowell, it also contains photographs, plat maps, artifacts, memorabilia, scrapbooks, gifts and even an old icebox, which is now displayed in the museum." The collection also includes forty-seven historic paintings of famous Americans associated with the Mississippi River.

Another person who can be found at the William Bowell Library is curator Tacie Campbell. For the past twenty-five years, Tacie has worked

Central English teacher Tim Ebeling and registrar Denise Vondran guide Central Shot Tower Investigators as they conduct research at the Captain William D. Bowell Sr. Library in the National Mississippi River Museum at the Port of Dubuque. *John Adelmann.*

on restoration and preservation projects. She has also completed a lot of research and has learned much information on local people and events through oral histories. When Tacie visited our school to meet with our Shot Tower research class, she gave us some information that helped us understand more about its early stages of design and construction.

Tacie told us, "The Dubuque Shot Tower was made in Egyptian Revival style of architecture," which means that its shape is squared off and angular, somewhat resembling the obelisks and pyramids in Egypt. Other shot towers across the country are cylinder-shaped and look a lot like smokestacks, and not many have the number of windows the Dubuque Shot Tower has.

The Bowell Library has many boxes of information about the different Dubuque history subjects that make doing research easier because it's all in one place. When it came to getting material on the Shot Tower, we used the photos, magazine articles and newspaper clippings we took from the boxes marked "Shot Tower." These were excellent sources for us. We were also allowed to make copies of some of these items so we could read them back at school, interpret them and then write our papers. Hopefully, one day our Shot Tower research book will be added to this impressive collection of information so others who come after us can find out what we discovered. It is really great to know that our class is making an important contribution to Dubuque's local history.

We are thankful for people like Denise and Tacie, and for the Bowell Library. Without them, Central students would have a more difficult time finding out accurate history on the Shot Tower.

Editor's note: In 2005 the William Bowell Library was located in the basement of the National Mississippi River Museum and Aquarium building. As the museum expanded, the library was moved into the old Diamond Jo Casino building that has since been renovated and is now an integral part of the museum complex at the Port of Dubuque.

Chapter 13

The Incredible Shrinking Shot Tower

"Focus on Youth" essay by Stephany Green

When our Shot Tower was built, the editor of the *Dubuque Express & Herald* reported on November 26, 1856, that it was 150 feet tall. In 2005, the *Dubuque Telegraph Herald* said that the tower was 117 feet tall. That's a 33-foot difference! What's going on here? While conducting my research trying to determine the true height of the Shot Tower, I've had the opportunity to interact with experts and private citizens who all have theories they think help explain the differences in the reporting of the Shot Tower's elevation.

I met Al Miller at the Center for Dubuque History at Loras College while our class was doing research there. When I mentioned that there was this huge difference in height in these two newspaper reports and that I couldn't figure out why, he simply said, "It sunk!" He believes this because the tower was built on an island that was nothing but sand and fill. But my question to that is: How could it sink straight down without tilting? And once it began to tilt a little, wouldn't it just keep going because it was so tall? However, while Al's theory is improbable, it is certainly not at all impossible.

Another idea about how tall the Shot Tower actually is comes from Jim Jacobsen, of Des Moines, who is a historic preservationist, consultant to the City of Dubuque and a guest speaker in our class at Central. He thinks that when the plans were made to build the Shot Tower, the designers planned to build it too high. They misjudged how they were going to do the tapering of the structure with stone, and as the walls got closer together, they had to turn to brick. But instead of stopping the taper, they kept it going, and by doing so they cut themselves short. If they had stopped the taper, they would have been able to build higher than they did. Jacobsen also says that the designers must have chosen the looks of the tower over the way it was going to function because they didn't stop the taper. But again, my question is: Where are the documents that show this is really what they did? And why would they decide to make the Shot Tower just look good if it wasn't going to work correctly?

Yet another theory is given by engineer Ron Balmer of IIW Engineering in Dubuque. He says that some thirty feet of the original tower is covered by landfill. He believes this because, well, where else could it have gone? The Shot Tower didn't just disappear or sink into the mud and remain perfectly straight. He also says the bottom windows are partially covered, and this suggests that dirt was spread around the tower for some reason and that eventually it covered up nearly thirty feet of the building. This sounds like the most logical explanation to me, especially when you consider that the bottom windows are partially covered.

I never thought researching the Shot Tower would wind up to be so frustrating. However, now that we're digging for answers, we'll keep working until we find them. In 1856, the Shot Tower cost $7,000 to build. Too bad they didn't keep better records on how they did it back then for those of us who want to know about it today.

Chapter 14
Fieldwork: Central Students Learn to Make Shot

"Focus on Youth" essay by Zac Brothers

The Helena Shot Tower in Spring Green, Wisconsin, and the Dubuque Shot Tower in Dubuque, Iowa, have a lot of similarities, such as how difficult it is to find someone who actually has information on the shot-making process at each location. We are learning about the Dubuque Tower, but we still wanted information about the Wisconsin tower, so our Shot Tower class traveled to Tower Hill State Park in Spring Green. When we arrived, park ranger Wayne Schutte gave us a tour and provided interesting facts about the history of the Helena Shot Tower and how it became a success.

We learned from Mr. Schutte that Green Bay businessman Daniel Whitney built the Helena Tower in 1831. It consisted of a smelter house that sat on the top of a cliff rising near the Wisconsin River's edge, a 180-foot shaft and a 90-foot horizontal tunnel to retrieve the shot that fell down the shaft and cooled in a pool of water. We also learned about shot-making. Lead was brought from Mineral Point, Wisconsin, and Galena, Illinois. The ore was then smelted in large kettles at the smelting house on top of the bluff. Arsenic was added to make the molten lead brittle to help it form into droplets when it was dropped down through the wooden shaft and through the hole dug into the stone bluff, all the way down to the pool of water to cool the lead.

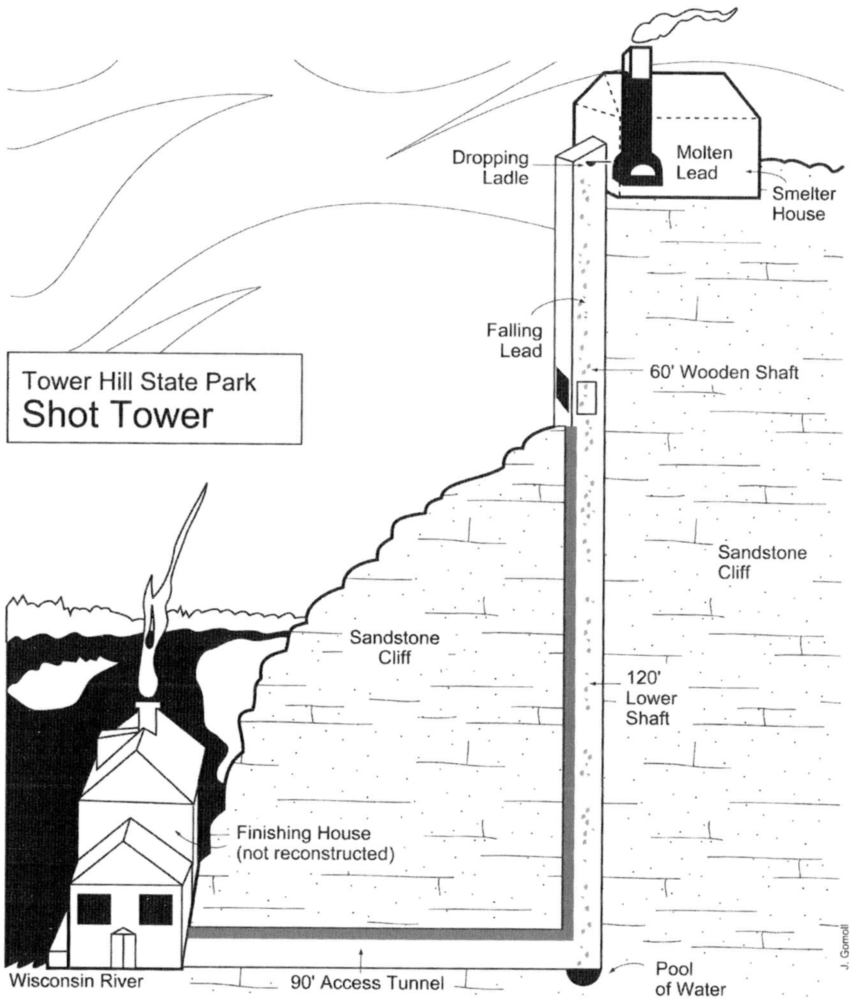

The Helena Shot Tower. The drop shaft was composed of an enclosed wooden structure built over a vertical tunnel chiseled into rock to create a 180-foot drop. Shot was retrieved via a horizontal tunnel. *Wisconsin Department of Natural Resources.*

Using a heavy ladle with holes in its lid, workers poured the hot liquid metal down the shaft, where it broke up into little droplets and cooled in the water below. Whitney and his employees dropped about five thousand pounds of lead per day, but only six to eight hundred pounds were usable for shot. Imperfect pellets were then hauled up from the river to the top of the cliff, melted down and poured again. We all got a chance to pour our own shot and then went down to the river's edge below the bluff and walked

Flicking the wrist caused molten lead to trickle through holes in this ladle's porous lid and fall to a tub of water below. Five thousand pounds of lead processed this way yielded about seven hundred pounds of useable shot. *John Adelmann.*

into the tunnel in the side of the cliff to see what the shot looked like in the pool of water. We all agreed that to build this Shot Tower took a lot of long, hard work, almost all of it done by hand with picks and shovels. It was also interesting to think that this tower was built about twenty-five years before the Dubuque Shot Tower.

Our class learned a few things that day—that wherever they are, shot towers are rare and should be more appreciated and that the shot-making business was very hard work, from the mining of the ore to getting enough firewood to melt the lead ore so it could be ready for the lead-dropping process, to sorting good shot from bad and then selling it for a profit. People should be informed of the important history that was made when people living in our area of the country decided to create shot ammunition using the lead they found here.

We thank Wayne Schutte for giving us a chance to go back into history, learn how shot was made and learn how shot-making became an important business here in the Midwest.

Chapter 15

A Sporting View of the Shot Tower

"Focus on Youth" essay by Spencer David

I like sports. I'm also researching and teaching myself about the Shot Tower. So I decided to combine these two interests to help me remember the information by using professional football. What do the Shot Tower and quarterbacks have in common? The answers might surprise you!

Quarterbacks are (1) expensive, (2) well known, (3) prone to injury, (4) can be traded and (5) can make comebacks. The more I thought about it, the more I realized the Shot Tower could be seen in the same light.

The George W. Rogers & Co. Shot Tower is (1) expensive. In 1856, the tower cost $7,000 to build. Today (2005), the City of Dubuque is planning to spend $600,000 just to restore the already standing tower, not even to build another one. The Shot Tower is (2) well known. Not only is it one of eight historic landmarks in Dubuque, but it's also on the registry of National Historic Sites. The Shot Tower was also (3) prone to injury. In 1911, a fire destroyed the observation deck and the wooden staircase inside it, and in 1991, a runaway railroad car hit the tower, doing $10,000 in damage. The Shot Tower was (4) "traded" to many owners. It started with George Rogers, who built it in 1856, who then sold it to Mr. Cook and Mr. Langworthy, who later leased it to Peleg Tallman and William Hyde Clark. When the Shot Tower was sold at auction in 1860, a man named Gilbert Woodruff put in the highest and only bid. Then J.K. Graves went to another auction and bid on the tower from Woodruff. Then in 1862, Graves turned it over to

Central Shot Tower Investigators Spencer David and Amanda Sproule meet with the Dubuque Questers to report on the progress of the Shot Tower research class. Civic groups and organizations frequently invited students to speak at their functions. *John Adelmann.*

St. Louis shot-makers Chadbourne & Forster. Wow! That's seven different owners in six years! The Shot Tower is (5) now making a comeback. Later in 2005, the City of Dubuque will use $600,000 to begin to restore it back to greatness, including the following improvements: (a) tuckpointing, (b) replacing windows, (c) rebuilding the staircase that leads to the top of the tower and (d) building another observation deck. Before long, the Shot Tower once again will be a famous and respected city landmark, just the way it should be. This football quarterback comparison helped me stay more focused and use my time wisely. I am also staying connected with the research because learning things this way makes it more fun to remember information on the Shot Tower's history.

Editor's note: For budgetary reasons, revised rehabilitation plans developed by the City of Dubuque postponed the installation of the Shot Tower's staircase and observation deck after this essay was written.

Chapter 16

Fieldwork: Learning Turns to Teaching

"Focus on Youth" essay by Zac Brothers

Central students are still researching the Dubuque Shot Tower. We've been writing articles for the *Telegraph Herald*, preparing speeches and working on the very first Shot Tower research book. We even took a field trip to see the Helena Shot Tower in Wisconsin. But we never thought we'd be asked to teach others about the Shot Tower.

One day, our class received a call from Mrs. Schreiber at Fulton Elementary School in Dubuque. Her third-grade students were learning about the history of the Shot Tower, and she wanted to get more facts from the Central students who were in the Shot Tower class. At first, we were excited, but then we realized that we were going to have to teach third graders, something none of us knew anything about. We're in high school—we don't know how to teach eight-year-olds, especially about a historical building such as the Shot Tower! However, after some discussion in class, Amanda Sproule, Spencer David and I volunteered to go to Fulton School and make a presentation to the students about the Shot Tower. We all wore our custom Central Shot Tower Investigators (CSI) T-shirts, but we didn't know what to expect when we got there.

Soon our fears turned to excitement again, because we came up with a plan of action. First, we introduced ourselves to break the ice. Second, we taught the students interesting facts about the Shot Tower. Third, for a physical activity, we had them pour water from a ladder through a strainer to form

CSI
Central
Shot Tower
Investigators

The Central Shot Tower Investigators T-shirt logo. *John Adelmann.*

into droplets to give the students the idea of how real molten lead was poured into a sieve and then allowed to fall into a cooling tub below. Everyone in the class wanted a chance to do some pouring. They got in line and took turns climbing up the ladder and dribbling the water through the sieve we made to a bucket on the floor. Finally, we gave them cardboard cutouts in the shape of one side of the Shot Tower so they could color in how they thought it looked. If four students joined up and taped their cardboard shapes together, they could create their own versions of the Shot Tower. This activity was really exciting for them. The students took this part of our lesson very seriously.

When we got back to Central and told our classmates how our teaching time went, Mandy Sproule said, "I was amazed by how well the Fulton students behaved and listened while we gave our presentation." Classmate Spencer David said, "I was even more amazed by how much they already knew about the Shot Tower."

After our presentation, Mrs. Schreiber sent us a note that said, "I wanted to let you know how much the Fulton third grade teachers and students enjoyed your presentation on the Shot Tower. Your enthusiasm for this monument and your research project was very evident." Then she said, "We would like to thank you for the additional effort and preparation you made to ensure that your presentation was appropriate for the third grade students."

We haven't just been learning about the Shot Tower, we've been teaching children as well as adults, and we plan to keep on doing it with our Shot Tower research book.

Chapter 17
An Interview with Historian Jim Jacobsen

"Focus on Youth" essay by Stephany Green

Our Shot Tower class met with important community people to discuss our Shot Tower project. Mr. Jim Jacobsen, a historic preservation consultant, was present. I asked him what we might encounter while researching our topic and why Dubuquers should be concerned enough to learn more about the Shot Tower.

"My main point is that it would be better if Dubuquers cared about the Shot Tower for the right reasons, which means basing the importance of the structure on facts and not myths," Jacobsen stated. He added, "They should also care about the Shot Tower because of its rarity, its engineering and design, its age and its link with the lead-mining industry that made Dubuque what it is today."

About challenges the Central Shot Tower researchers may encounter, Jim said, "There are a lot more myths and legends than facts about the Shot Tower, and the places left to look for answers are the very hardest to reach. For example, the Shot Tower's archaeology—how it was built—will answer design questions. Why hasn't it fallen down long ago? What's holding it up?"

Jim's consulting business is called History Pays, so I asked him, how is history going to "pay" for those of us doing the research and the members of the Dubuque community who will be reading it? Jim replied, "History is a toolbox we can get ideas and wisdom from to learn how people in the

Des Moines historical consultant Jim Jacobsen shares his theory of how the Shot Tower was built with the Central Shot Tower Investigators during a class presentation. His groundbreaking research became an invaluable teaching and learning tool. *John Adelmann.*

past solved their problems, or how they improved their communities. Also, history reminds us that every place is unique and important. History consists of what has been, what is and what is going to be. The people who came before us already gave it their best shot, and we inherited their architecture, their businesses, their industries, their governments and their libraries. The smart thing to do is take advantage of all of the life experiences that they lived through and learn from their mistakes and successes."

When I started this project, I didn't know what a shot tower was and didn't care about history, but Jim's words lit a fire in me. I now understand that there is meaning to how things used to be and that history is not just something that happened a long time ago that doesn't affect me. Today, my classmates and I look forward to working on the Shot Tower project to preserve the past for future generations.

Chapter 18

The George W. Rogers & Co. Shot Tower: Other Perspectives

By James E. Jacobsen

As a historic preservation consultant with a business called History Pays, it was my privilege to be the first historian to complete a comprehensive study of the George W. Rogers & Co. Shot Tower for the City of Dubuque. It was a daunting task, given the fact that the tower was an icon for Dubuque, if not for Iowa. An icon, by my application of the term, means that the structure was more valued for its myths than for its actual history. And like all icons, its sheer survival against all odds for over 150 years conferred upon itself the status of a landmark.

Few shot towers survive anywhere else in the country today, so the fact that the one in Dubuque still stands is quite remarkable. But there is another level of understanding and appreciation that most visitors miss. Similar to an iceberg, there is almost as much of the Shot Tower's structure that is underground and invisible as there is above ground and observable. Dubuque's Shot Tower was built in the middle of nowhere on the middle of nothing. There is no reachable bedrock where it stands, at least in mid-nineteenth-century terms. So what supportive superstructure accounts for the fact that the Dubuque Shot Tower currently is as plumb and level as the day it was finished? The answer rests in the engineering of similar structures, such as chimneys and water towers. They were constructed with angled wooden pilings arrayed in a cross-pattern, with multiple rows of pilings, those in the center being driven straight down, while those on either side were positioned at increasingly splayed angles. Atop these was a solid base of heavy squared timbers, and atop that was a limestone pyramid. The presence of the wooden superstructure was detected during the archaeological probes done in 2006 (see Chapter 41). The wooden components remain solid because they are set beneath the water table. All in all, it is a remarkable engineering feat—designing and constructing this building on a mud flat in the middle of the Mississippi River.

Which leads to the next point. Why is the Shot Tower where it is? Even today, it takes a long trek to get to it, and at the time it was built, the city proper was much farther distant to the southwest; there was nothing but canals and wetlands between the Shot Tower and town. All shot towers were located at shipping points, which meant by a body of water. No other ingredient—the lead ingots, the labor, the coal or wood necessary for the furnace—was readily at hand. The only convenience was the availability of a well of cold river water at the tower's base. The archaeologist confirmed the Historic American Buildings (HABS) drawings from the 1930s. The tower's base is a hollow, water-filled basement where the hot lead droplets could instantly finish their cooling descent.

Another related question: why was the Shot Tower located in Dubuque? Naturally, because of the availability of locally mined and smelted lead. But the structure also represents a rather belated realization by the citizens of the town that when "value" was added to local products, it meant more revenue for Dubuque. Put another way, smelted and processed lead was worth more on the open market than just mined lead ore. Similarly, finished lead products also had proportionally higher values, and they could be shipped for longer distances at a profit. Thus, the Dubuque Shot Tower could compete successfully with St. Louis shot producers whose costs to ship their products upstream were more than those for the businessmen in Dubuque who shipped their goods downstream. Until the Shot Tower was built, Dubuquers had to purchase shot from New York via St. Louis at considerably higher prices.

The research also reveals that the actual shot production history of the Dubuque Shot Tower was less continuous than legend would have everyone believe. And yes, Virginia, there were no cannonballs dropped in the Dubuque Shot Tower! Once again, the tower's symbol of shot-making would seem to be more important to some than its actual historical function. And when considering symbols, it is important to realize two things. First, the Shot Tower is made in the shape of an Egyptian obelisk. Second, it was never really finished. The obelisk form is endemic to the structure type, which is to say almost all shot towers had four narrowing sides. Like all tapered load-bearing structures, less wall is required as the construction goes upward. The obelisk's notable shape, like the Washington Monument, was the national form for monuments, and it was particularly popular in Mississippi River cities. And because at that time the Mississippi River was referred to as "America's Nile," it appears that the Dubuque Jail—built next to the Dubuque County Courthouse within a year of the Shot Tower in 1857, sporting the Egyptian Revival style of architecture—and the Dubuque Shot Tower were conscious imitations of the structures and designs that Napoleon's scientists and sailors were stealing from the real Nile River Valley in the very early part of that same century.

The "unfinished tower" concept is my idea completely, and there is not a scintilla of historical evidence for the notion so far. Recalling that the designer and builder of the Dubuque Shot Tower remain unknown (although some have their suspicions), the second reason for placing it on the riverfront was to "boom" Dubuque by giving it a prominent monument. It was also supposed to be a focal point for real estate promotion to generate sales of lots on the newly created land made possible by the Dubuque Harbor Improvement Company at the northern end of the city's riverfront. However, absent any architectural plans, there is no explanation for the combined use of both limestone on the lower section and brick on the upper section of the Shot Tower. We do know that the limestone came from Illinois and not Dubuque. That was convenient. We understand that the higher, thinner walls could better be constructed using brick rather than stone.

We also know from newspaper accounts of the day that at one point, the Shot Tower was to have included a light source of some sort on its roof to help aid in river navigation. All of this evidence tells me that the original builder envisioned this particular shot tower as a massive Egyptian obelisk, complete with a spotlight to aid traveling steamboats up and down this section of the river. An application of stucco would have rendered the whole structure in its true, smoothly completed form. But unfinished, it would live out its years as we have always known it.

We also know that the tower's construction coincided with a national over-investment in all things, particularly in railroads. The ensuing economic downturn, termed the Panic of 1857–58, put some local Shot Tower owners in a difficult financial situation and left the rest of the country in economic doldrums that only a civil war would end. The Shot Tower was also a victim as it was sold and resold, used and then unused—actually for a longer time than when it was in operation turning local lead ore into finished shot for market. But at the end of the day, the George W. Rogers & Co. Shot Tower in Dubuque, Iowa, is still an important historic symbol and reminder of the entrepreneurial spirit that was alive and well in America in the mid-nineteenth century.

Chapter 19

John Francis Rague: Pioneer Architect of Iowa—and the Shot Tower?

By Chris Steines

Editor's note: Information used in this chapter comes from an article entitled "John Francis Rague—Pioneer Architect of Iowa," written by M.M. Hoffman in the Annals of Iowa *19, no. 6, published in October 1934.*

Architect John Francis Rague is referred to by some historians as the "pioneer architect of Iowa." He is well known for designing and building structures in various states and in the Midwest. Some of these states include Wisconsin, Illinois, New York and Iowa. His name is linked to the Shot Tower because of certain circumstances. The design style he used to create one particular building in Dubuque suggests he might have had something to do with the design and construction of the George W. Rogers & Co. Shot Tower.

Rague's father, Dr. John Rague, served as a surgeon and personal physician to General Marquis de Lafayette in the French army, and he came to America with the general to fight on the side of the American colonies during the Revolutionary War. After that conflict ended, he remained in America. The birth record states the following about his son John: "John Francis Rague, born at Scotch Plains, New Jersey, on March 24, 1799." Rague's mother was forty-one when she gave birth to her son, and she had been a friend of the mother of Washington Irving, the famous writer. She also took care of the young Irving child when his mother became ill. In 1804, the Rague family moved to New York City.

In 1806, a young John Rague attended school in New York and then later got his degree in architecture by studying with the famous designer and builder Milard Le Fevre. Rague also met Eliza Vandyke, whom he courted and later married. A few years later, in 1824, General Lafayette returned to America for "his last triumphal visit to America" and found out that the son of his personal physician, who had died some time before, was "in the assemblage" at an event organized in his honor. Hearing this news, General Lafayette asked that John Rague be escorted to the head table and made arrangements to have him sit with all the special guests. So in his early life, John Rague learned to become an architect from another famous architect and also had friends in very high places.

After a few years of working in New York, Rague moved west to live in the growing town of Springfield, Illinois, in 1831. In Springfield, Rague got involved with the Presbyterian Church, and he joined the choir. He was known for being "a musician with more than ordinary talents and endowed with a rare tenor voice, he became a leader in the church choir." While in the choir, he met some soon-to-be well-known people in American history, including Stephen A. Douglas, Mary Todd and Abraham Lincoln.

Author M.M. Hoffman states that John Rague and Abraham Lincoln became good friends, even though John was ten years older than Abe. The "awkward young lawyer" Lincoln asked Rague to show him how to dress for public functions and "allowed himself to be groomed" by the polished and sophisticated architect who came from New York. In fact, Abe Lincoln decided to wear white gloves for the first time to attend a dance after he was encouraged by John Rague to do so. This piece of history comes from an interview conducted by M.M. Hoffman in 1928 with Mrs. John O'Keefe, who was a business partner and companion to "the second Mrs. Rague." She had in her possession the Rague family Bible, "in which is written down much concerning Mr. Rague."

In 1836, the people of Springfield elected Rague to be a town trustee. During the following year, Springfield became the capital of Illinois, and John Rague was asked to design the new statehouse at a salary of $1,000 a year. The structure would be in Greek Revival–style design of architecture. According to the *Biographical Dictionary of Iowa*, this building was completed "between 1837 and 1853." The design and construction of the Illinois statehouse in Springfield brought Rague some additional fame. He was then asked to design the first capitol building of Iowa, located in Iowa City, in 1839. This building would also be built in the Greek Revival style. Its cornerstone was laid on July 4, 1840, and lawmakers began to use it in 1842. However, this building took an additional thirteen years to complete. Rague designed the plans for that building while he was still living in Springfield.

In 1844, Rague traveled to the city of Milwaukee. It is interesting to note that as a resident in this area, he did not refer to any of the success he had in Illinois or Iowa to get new contracts for business. In his advertisements in the newspaper he simply described himself as follows: "After twenty years of practical building in the city of New York, he will draw plans and specifications and contracts for all types of buildings." Rague was one of the first architects to work in Milwaukee. There, he designed the Phoenix Building and several school buildings. He also spent a lot of time in the cities of Chicago, Illinois, and Madison and Janesville, located in Wisconsin.

Besides being an architect, Rague was also interested in other things, including music and politics. Rague became treasurer of the first music organization in Milwaukee, the Beethoven Society. In 1846, he failed to win an election for justice of the peace, and three years later he was defeated in an alderman race. But his skill as a designer of buildings was known to all. For example, in describing Rague's architectural work in Madison, Wisconsin, a man by the name of Arthur Peabody wrote the following very kind description of his abilities that can be found in the *Wisconsin Magazine of History* 10 (December 1926):

> *The most interesting record of the man concerns the designing of the three buildings for the University of Wisconsin: University Hall (now called Bascom Hall), North Hall, and South Hall…The buildings still remain and have been admired by several architects of note for their simple and refined architectural character. It would be a graceful thing to inscribe his name on these buildings. The records of the Board of Regents of 1859 and the notices of the Wisconsin Argus of the time are all that an architect could desire for recommendation.*

The *Biographical Dictionary of Iowa* states that Rague's wife, Eliza, divorced him in 1851 but does not explain the reasons for this. He later met a woman in Janesville named Chestina Scales. Chestina's family were strict members of the Episcopal Church and wouldn't allow her to marry a divorced man. However, she and Rague were later married anyway, and they moved to Dubuque, Iowa, in 1854.

In Dubuque, Rague met Stephen Hempstead, a man who had just returned from Iowa City after serving four years as governor of the state

Stephen P. Hempstead (1812–1883). Iowa's second governor from 1850 to 1854, Hempstead encouraged his friend and (Shot Tower?) architect John Rague and his wife to permanently settle in Dubuque after they arrived in 1854. *Dubuque County Historical Society.*

of Iowa. Rague and Hempstead became good friends. Hempstead persuaded Rague to become a permanent resident of the city of Dubuque and Iowa. This relationship seemed to help Rague get work in Dubuque, but perhaps it was because he was just well known all by himself and didn't really need the governor's help.

Rague was elected to the Dubuque school board and became an active member. Once here in town, he also designed Dubuque City Hall, the Octagon House (also known as the Langworthy House), the Mathias Ham House and schoolhouses in the first, third and fifth wards. One other important building Rague designed was the city's first jail, what is now called the Old Jail, a building that is now used as a local art gallery and museum. This structure was made in a design called Egyptian Revival style. According to author Hoffman, Rague modeled this building "closely after the old Tombs of New York, even down to the fierce, medieval-looking dungeons deep below the structure." The jail was later described by James O'Donnell Bennett, the literary and cultural critic of the *Chicago Tribune*, as "this gem of Egyptian architecture, transplanted across the Mississippi River."

In 1856, the George W. Rogers & Co. Shot Tower was built in Dubuque. We know that John Rague was living here at this time. Des Moines historian Jim Jacobsen notes in the research that he did for the City of Dubuque that the Shot Tower is very much Rague's type of Egyptian Revival style of architecture. Many shot towers built in other parts of the country looked more like tall smokestacks, and most of them didn't have any windows. However, Dubuque's Shot Tower has squared-off, flat sides with many windows and has a taper to it. This leads some to believe that Rague may have been the person responsible for its design because there are interesting

The Octagon House, also called the Langworthy House, was designed in 1856 by John Rague for miner and businessman James Langworthy. This noted Dubuque residence has remained in the Langworthy family for five generations. *Jack Boucher, National Park Service.*

Dubuque's first city jail was designed by John Rague and built in 1857 in the Egyptian Revival style of architecture. Some researchers believe the Shot Tower's unique obelisk-like shape also reflects this distinctive motif. *Center for Dubuque History.*

similarities between the Egyptian Revival style of the Old Jail and the shape and design of the Shot Tower, which looks a lot like an ancient Egyptian obelisk. In addition, Jacobsen says that during these years the Mississippi River was also called "America's Nile," so it would seem that a shot tower built to look like an Egyptian obelisk on the bank of the Mississippi could have been Rague's idea.

After doing research at Loras College, I found that George W. Rogers, gunsmith, and John F. Rague, architect, actually had offices in downtown Dubuque not very far from each other. In fact, I found two classified ads in an old newspaper, one on top of the other. One was for "John F. Rague, Architect & Superintendent of Buildings." His office was listed as being on Main Street, "opposite Barney & Co.'s Bank." Right underneath this was an ad for "G.W. Rogers & Co., manufacturer of Patent Shot." Customers were told that his business could be found at the "Levee, Dubuque, Iowa," in case they might not have been able to locate the Shot Tower itself! Chances are these two businessmen knew each other, but we are not able to prove it.

Even though he was a well-known and successful man, tragedy attacked John Rague. In the 1860s, his eyesight began to fade, and a few years later he became completely blind. Stephen Hempstead also had a tragic accident and his right leg was amputated. As a result, the two friends no longer visited with each other. Eventually, Rague and his first wife, Eliza, reconciled their differences, and she came back to visit him several times "in his affliction," meaning as he was going blind. She later died of natural causes. Her body was brought to Dubuque at the request of Rague and was buried in Linwood Cemetery.

John Francis Rague arranged his own funeral ceremony before he passed away on September 24, 1877. He also was buried in Linwood, along with Eliza and second wife Chestina. Before he died, he wrote an interesting epitaph that he wanted to be inscribed on his tombstone:

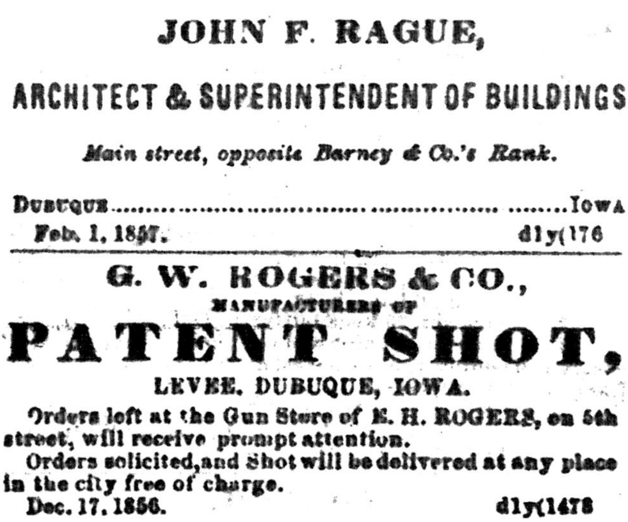
JOHN F. RAGUE,

ARCHITECT & SUPERINTENDENT OF BUILDINGS

Main street, opposite Barney & Co.'s Bank.

DUBUQUE .... IOWA

Feb. 1, 1857. d1y(176

G. W. ROGERS & CO.,

MANUFACTURERS OF

PATENT SHOT,

LEVEE, DUBUQUE, IOWA.

Orders left at the Gun Store of E. H. ROGERS, on 5th street, will receive prompt attention.

Orders solicited, and Shot will be delivered at any place in the city free of charge.

Dec. 17, 1856. d1y(1478

Classified ads taken out by John Rague and George W. Rogers reveal that their businesses were located within blocks of each other, suggesting these men might have known one another and collaborated to build the Shot Tower. *Dubuque Express & Herald.*

This planet earth, its face I've trod
For three score years and o'er,
Now its bosom make my bed,
To rest for evermore.

Through ere a thousand years shall pass,
My dust shall rise again;
May generate e'en flowers or grass,
By aid of sun and rain.

The bees will sip these fragrant flowers,
The lambs will eat the grass,
And thus they'll spend their earthly hours,
Till from this life they pass.

Then all return to mother earth,
Some time again to rise,
Though no one knows the kind of birth,
But God, the only wise.

Thus Nature's laws are God's cause,
Obedient to his will,
Men sometimes teach, but let them pause,
All Nature speaks—be still.

Roll on our planet, in the train
With million others, roll.
Man need not fear, he can't be slain—
He's under God's control.

Although it is not proven beyond a doubt that this talented man was the architect who designed the George W. Rogers & Co. Shot Tower, the finger of history and the designs he left behind all point to him. John Francis Rague, known as the "Pioneer Architect of Iowa," will go down in America's history for his brilliant work of designing state capitols, buildings, schools and homes in the Midwest and also in Dubuque, Iowa. Just maybe he also could be known for designing the old structure we now call the Shot Tower on the banks of the Mississippi River.

Chapter 20
The Dubuque Shot Tower in Action

By Jake Christensen

Having read this far, readers of this book should realize that information about the George W. Rogers & Co. Shot Tower is not to be found in textbooks. Of course, we knew that there was information out there—we just had to find it, read it and analyze it. We were fortunate enough to find a nearby place where microfilm is used to store old local newspapers that were published at the time of the Shot Tower's planning, construction and operation. The location of the microfilm we used was at the Center for Dubuque History at Loras College.

The historical information I used in this chapter comes directly from the old newspapers of those days. As closely as possible, I have tried to define certain words and explain quotes and break them down to make sure you notice everything of significance. In other cases, the quotations and news accounts will speak for themselves. Let us now turn back the pages of time to explore the history of Dubuque's Shot Tower and see what people 150 years ago actually read while it was becoming a reality and then, once it was built, how it operated. Spelling and punctuation that were used back then are shown here as well.

In 1854, a man named George W. Rogers settled in Dubuque and opened up a gunsmith shop. His cockiness can be seen in the ad he ran in the local paper that "declared war" on all the other businesses that made guns. However, it would not be long before Mr. Rogers would become known for

WAR DECLARED.

THE subscriber having permanently settled in Dubuque, would respectfully say to the citizens and the inhabitants of the inhabitants of the North and West, that he is now prepared to make to order or furnish Rifles of every description; Patent Muzzle Telescope Rifles, with increase twist; also, all kinds of single and double Rifles, with prices and finish to suit all. He is is now the agent for Sharp's Patent Breach Loading Rifle, which can be loaded and fired with the greatest accuracy ten times per minute, which for its great accuracy, convenience and durability, is celebrated throughout the world. I am also agent for the Revolving Hammer Pistol, which may be seen at my shop. Also, Colt's, and Allen's, Revolvers, Dueling and other PISTOLS. Double and Single Guns of every description, always on hand. Powder, Caps, Lead, Shot, Mine Cartridge, Ely's Water Proof Caps; Also, United States Caps of a superior quality, Flasks, Shot Belts, Pouches, Pistol Flasks, and every thing usually kept in the line.

Those who are in want of any of the above articles are particularly invited to call and examine my Guns, which for durability and accuracy, I challenge the world.

Gentlemen can be furnished with Guns by the day on reasonable terms.

☞ All kinds of Repairing done with promptness and dispatch, and warranted. Thankful for past favors, I respectfully solicit a continuance of the same.

Shop on Fifth street, below Main street, a few doors from the Globe Building, sign of the Elk Horns.

G. W. ROGERS.

Dubuque, Dec. 23. 1854. dawtf(ch

Gunsmith George W. Rogers announces the opening of his business in Dubuque by "declaring war" on other local gun dealers in December 1854. Rogers would begin construction of the Shot Tower less than two years later. *Dubuque Express & Herald.*

more than just making pistols and rifles. On December 3, 1855, the *Dubuque Daily Express & Herald* mentioned that someone was at least thinking about building a shot tower in Dubuque to take advantage of all the lead that was being mined in the area and had taken the first step to get it done by purchasing some land: "A Shot Tower in Dubuque—Mr. G.W. Rogers has had it in contemplation, for some time, to erect a shot tower in this city, and we learn that he has now purchased the ground and is making preparation to build. He has made an improvement in the method by which he greatly reduces the expense."

Notice how the article doesn't explain how George W. Rogers will "greatly reduce his expense." Just two months later, the first indication that a business was going to be created that had something to do with manufacturing lead shot was found in an article in the *Express & Herald* of February 15, 1856, which read: "G.W. Rogers of this city, has purchased a lot in the center of the island, on which he intends to erect immediately a shot tower, on the top of which the Dubuque Harbor Improvement Company design to place a brilliant white light, during the season of navigation, for the benefit of steamboats landing at night, and to light up the levee." There was no other mention of what that "brilliant white light" would be. Perhaps a fire? It seems this would have been a hard thing to keep going on top of such a small place like the top of the tower.

About eight months later, on August 2, 1856, in an article in the *Dubuque Daily Republican*, we read that the planned construction of the Shot Tower is now on approach. Notice how the site where the tower is to be constructed is described, and also note the structure's intended height:

> *GRAND PROJECT—We learn that Messrs. G.W. Rogers & Co., Manufactures of Guns, Pistols, and general dealers in hunting apparatus, & cutlery, of this city, will soon erect a Shot Tower on the Island, at the foot of 7th Street, near the new Levee. It will be built of stone, 25 feet at the base, and ten feet at the top. The Tower will be 150 feet in height from the base, and will be capable of manufacturing three to five tons of shot of all sizes, and bar lead, per day. It will be worked by a five horsepower steam engine.*

Notice how in this next section of the article the reporter seems to boast about the Dubuque Shot Tower's superiority over the rival St. Louis shot-making company of Chadbourne & Forster: "The amount of capital invested is from $10,000 to $12,000. It is intended that the Tower shall be completed and in successful operation about the 1st of October next. Merchants of Chicago and the great West, will find Dubuque the best market for shot and lead. This city has a decided superiority over St. Louis."

When the local paper boasted about the tower, it seemed to be choosing sides as well. It isn't understood why the articles did this because press today isn't able to talk like this and take sides, except in editorials. The Shot Tower's owner also had a gun-making business, so making shot at the tower would be just one more thing he could sell to his gun customers to make a profit. And his shop wouldn't be all that far away from the riverbank where the Shot Tower stood:

> *Being so favorably situated, surrounded by hills and bluffs abounding with an inexhaustible supply of mineral, and such immense shipping resources must at once determine this to be the great lead mart of the Union. This enterprising firm intends to carry on the manufacture of every description*

> *of guns, also keep a general assortment of revolvers, of the most approved patents, fine cutlery, and materials for dealers. Their establishment is on 5th Street, between Iowa and Main.*

When the article said "this [is] to be the great lead mart of the Union," it sounded as if the paper was saying there is no better place to build a shot tower in the entire country because all the natural resources and other factors that go into creating shot are right here in Dubuque.

On September 27, 1856, the *Daily Republican* announced that the Shot Tower was now under construction. The height of 150 feet is reported here as it was in other papers: "The Shot Tower of Messrs. Rogers & Co. now in the process of erection, on the levee near the foot of 7th Street, is rapidly approaching completion. It has now reached the height of 95 feet. It is being built of stone, and the extreme height of the tower will be 150 feet when completed."

In an article published on October 10, 1856, the *Daily Republican* made specific and detailed note of the tower's dimensions, suggesting it would indeed be gigantic:

> *Shot Tower.—Messrs. G.W. Rogers & Co. are building a large Shot Tower near the foot of 7th Street on the outer levee. It will, when completed, reach to the height of 150 feet, being 25 feet square at the base and 10 feet at the top, 105 feet high will be built of stone, and the remaining 45 feet of brick. This when completed will present quite an imposing appearance.*

In the same article we read once again that the owners of the tower were looking into expanding its uses and that they were not only for shot-making: "This enterprising firm have some intention of placing a large light at the top for the convenience of those navigating the river, if the city should deem it expedient." The article also boasted about the "light" on top of the tower being favored by ship captains and stressed the fact that the required lead ore needed to make shot could be found right here in our own backyard:

> *It certainly would meet with approbation* [approval] *of steamboat men. No city in the Union possesses better facilities for the manufacturing of shot of all sizes and bar lead than Dubuque, being surrounded, as it is, by high bluffs, containing an inexhaustible supply of mineral. The mines in this vicinity are very extensive, and the amount of lead ore annually shipped*

> *from them is immense. Dealers will find far better inducements to purchase here than elsewhere. This extensive establishment will be driven by a five horse power steam engine. The entire cost will amount to $12,000.*

Updated information on the Shot Tower, reported by the *Daily Republican* on November 12, 1856, gave people who read the paper a good indication that the Shot Tower's construction would soon be finished:

> *We learn that the Shot Tower of Messrs. G.W. Rogers & Co., at the foot of 7th Street on the outer Levee, will be ready for business in a few days. It is now nearly completed, and ere long we shall hear of the Dubuque Shot Tower being in successful operation. We have no doubt the worthy proprietors will receive a highly profitable trade. Success to them.*

On November 18, 1856, the *Express & Herald* described the first-ever shot produced at the new tower: "The First Shot: We had the extreme gratification last evening to witness the first practical results of the working of Messrs. Geo. W. Rogers & Co.'s Shot Tower." Then the article explained how Shot Tower builder and owner George W. Rogers made sure everyone would know what kind of quality shot he was now making down by the Mississippi River. Rogers actually brought some samples as proof to the offices where the newspaper was published:

> *The shot placed upon our table by Mr. Rogers were unfinished and in the precise condition in which they were made—numerous sizes mingle together, yet the great majority very perfect, indeed, this first result must be very flattering to the proprietors, and unless we are much mistaken, this, which marks a new era in Dubuque manufactures* [sic], *will eventually, possessing the advantages of raw material at hand, open a very extensive trade, and one which cannot but prove a realization of a fortune to the parties interested.*

Once again, observe how the newspaper article appears to boast about the Shot Tower business now getting ready in Dubuque and that it was sure to be successful. Was the role of the newspaper back then to become an advertising agency for one man's company?

Now imagine you are back at the time of the new Shot Tower on November 26, 1856. It's a brisk, sunny morning as we get set to take a walk.

An unsigned and undated oil painting depicting the Shot Tower's proximity to the Mississippi River and distance from the town. Note the riverboat moored alongside the tower and the narrow pedestrian bridge that spans the sloughs. *Center for Dubuque History.*

Our journey is leading us in the direction of the George W. Rogers & Co. Shot Tower, and I shall take you on a tour of the historic building as it was described in the *Express & Herald.* The writer of the article began by saying, "As we walk the extension of 7th Street we can see the tall structure of the newly constructed Dubuque Shot Tower." Notice what kind of machinery is present in the tower and how it's all placed in specific locations for a reason. Not only is there a method for getting the lead up to the top of the tower, but there is also a process for finishing and shining, or polishing, the completed shot. The special treat is that we will also be able to meet the man who made it all possible. The paper reported it this way:

> *After taking a momentary glance we walked up the first flight of stairs, which makes an elevation of 10 feet, and entered the first story. In this room or story, will be found directly in front of the entrance, the steam engine which is used to hoist, by means of a windlass, the lead to the extreme height, preparatory to casting, and to rotate the cask used in finishing the shot.*

The reporter then described the area where shot dropped from the top of the tower and was caught and finished: "Besides this, there is the tub, partially filled with water, in the center of the chamber, in which is caught the falling shot; the right of the room on entering is occupied by the windlass and the left the inclined plane which detects the perfect from imperfect shot."

Then the reporter made his way up to the top of the Shot Tower to see how shot was made and met up with the owner of the tower himself, George W. Rogers, who was busy producing shot and trying to make his methods better: "After satisfying our curiosity on the bottom floor, we commenced the ascent, the means used, temporary ladders, and these we climbed from story to story, until we stood in the ninth, where we found one of the gentlemanly proprietors, Mr. Geo. W. Rogers, busily engaged in improving his furnace."

Now at the very top of the Dubuque Shot Tower, the reporter provided his readers with great detail about how shot was produced. He described the tools and how they were used:

> *This ninth story, which is just one hundred and forty feet from the ground (the height on the ninth floor), is where the great mystery is solved how they manufacture shot. The apparatus used, and of which consists the whole of this room's furnishings, is summoned up in a furnace capable of smelting three hundred pounds of prepared lead, and a ladle or two, together with the various sieves used in making the several grades of shot.*

Then the reporter described the mixture that was needed to make the lead workable and remembered to tell readers that each floor in the Shot Tower had a large hole, or aperture, cut out of it so the lead could fall freely from the top story. He then went on to describe the process:

> *After having explained to us how, that by taking a ladle full of the combination of arsenic and lead, heated to a certain temperature, and pouring the same into the sieve which is simply a sheet iron disk with small holes punctured at intervals of a half-inch, which hangs over the aperture, through the floor—and by the by, we should have mentioned that in the several floors to the very bottom there are apertures in each, thus forming a channel for the falling globules to pass—and from thence the molten lead passes to a tub below.*

After the shot was dropped and caught in the water tub below, the next process of how the good shot was separated from the bad and how the shot was graded was described in one very long sentence:

> *How that from this tub the shot is taken and placed in the cask, which revolves around until it receives that finish which is observed on all shot, and how it is taken from this cask and placed upon the inclined plane where the perfect shot roll down and the imperfect, oval and flat, slide down and drop into a groove which the brisk, round perfect shot leap over and are caught into a receiver, and how from thence they are passed through many sieves as there are sizes of shot, which are placed in a case one above the other, the coarser on top, and are thus graded preparatory to bagging and how they are bagged, ready for market.*

The reporter then took the last part of the journey and climbed onto the top of the Shot Tower and described what he saw:

> *After explaining all this to us, we placed our feet on the rounds of the ladder and emerged once more into the open air through the scuttle of the roof—and what a glorious sight was presented to our eyes—the whole city lit up with joyous sunshine looking like a fairy picture. Down the river, a view of eight miles with a boat in the distance steaming the river, and above, the glad waters shut out by the bluffs of the Wisconsin side, joining in appearance with that of Eagle Point on this, all presented, with the moving forms of life pursuing their avocations, and the hum of business which came from the work shops, a scene exhilarating in the extreme and worth the toil of climbing the height of one hundred and fifty feet to enjoy it.*

On the day before Christmas 1856, George Rogers ran an advertisement in the *Express & Herald* that announced he was finally ready to open for business and that he would be able to fill all orders for shot. This ad was supposed to be for local customers. Also, Rogers made it clear that he would not take any checks or credit cards:

> *Shot Tower—the subscribers are now prepared to fill all orders for SHOT* [sic]*, combining all the different sizes from Buck Shot to No. 20, at nearly St. Louis prices. All orders will be promptly filled, and shot delivered without any extra expense. Orders from the country must be addressed to*

> *G.W. Rogers & Co., accompanied by the cash, to receive attention. As our business will be strictly a cash business, we shall be enabled to serve those who may favor us with their orders at rates that will make it in their interest to deal with us in preference to other markets. Orders left at the Gun Shop of E.H. Rogers, on Fifth below Main Street, will receive immediate attention.*

Researchers believe that E.H. Rogers was George's brother who had taken over the running of the gun shop so George could concentrate on building and operating the Shot Tower. Back in August 1856, when George Rogers made the change from gunsmith to shot-maker, E.H. Rogers had placed an ad in the paper that announced he would now be running the store.

By 1857, the Shot Tower was running and producing good quantities of shot. On September 22 of that year, the editor of the *Daily Northwest* also made a visit to the Shot Tower, and what he described certainly helps us understand how shot was made back in the old days. It is also interesting to note that by this time, George W. Rogers was no longer operating the tower. In charge now were James Langworthy and Peter Cook. The editor wrote, "We chanced to pass near the Shot Tower, owned by Messrs. Cook and Langworthy. And as usual, our curiosity being excited, we determined to know and see how shot were made, so in we went, putting, as a matter of course, our best foot foremost." Here the editor explained that Mr. Langworthy had offered

HAVING purchased the interest of G. W. Rogers in the Gun business, I will continue it at the old stand, on Fifth street, below Main street, at the sign of the big gun, where may be found a large addition to the former stock, which consists in part of the following articles, viz:

Rifles, of my own manufacture and Eastern make.
Double Barreled Rifle and Shot Guns,
Single Barreled Shot Guns and Pistols,
Colt's, Allens'. Maynard's and Warner's Repeaters.
Also, the Volcanic Repeaters.
Deane & Adams' English Repeaters.
Robbins' and Lawrence Co.'s Revolving Hammer Pistols.
Powder in cannisters, and by retail, all varieties.
Shot Bags, Game Bags, Powder Flasks, Gun Wads and Gun implements.
Fishing Tackles, Pocket and Pen Knives, Bowie Knives, best assortment in the city.
Razors, Ely's wire Cartridges, Saloon Pistols, and caps for the same.

☞Repairing of Guns done Promptly, and warranted.

☞being Agent for Hitchcock & Mussy, Rifle Barrel makers, I will keep a large supply always on hand, for Gun makers, and all things appertaining to the trade, as low as any city west of New York.

Twenty years experience in the business enables me to say that, that all work entrusted to my care shall be as well done as at any similar establishment in the United States.

E. H. ROGERS.

Dubuque, Aug. 20, 1856.—dly(1018

E.H. Rogers publicizes the purchase of his brother's gun-making enterprise in August 1856. Historians believe George Rogers was now focused on completing the Shot Tower, the manufactured product of which would certainly complement his firearms business. *Dubuque Express & Herald.*

him a free tour of the facility. Even though he was getting tired by the time he had reached the halfway point, this newspaperman was determined to keep going until he reached the top of the Shot Tower:

> *By enquiry of Mr. Langworthy, one of the proprietors, we ascertained that anyone ambitious of going to the top of the tower could do so without charge, and free-gratis-for-nothing. So up we went,* puffing *and blowing, (that's our forte), until about midway, when we came to a halt, pretty well winded, and a strong inclination to turn back. But no—that wouldn't exactly prove our pluck and satisfy our* rising *ambition, so to the ladder we again returned.*

Then the editor revealed that it took a little sip of liquor to help him get over the fact that he was so tired from the straining climb to the top of the tower:

> *At an expense of much breath, and considerable straining of our—just now relaxed muscles—we reached the top, which is one hundred and fifty feet high. After resting a spell, and feeling somewhat recovered, and greasing our tongue with the "oil of civility," we commenced our inquiries of the presiding genius of the furnace, ladle and kettle.*

Perhaps Mr. Langworthy was a little puzzled with this nosy newspaper editor who was now out of breath and sipping on booze to help him get over it, but he kept on his task, and the owner of the Shot Tower was a gentleman. The article continued, "He [Langworthy] appeared somewhat bemused at our questions, all of which, however, he very politely answered. For the information of those who do not know how shot are made, we will give an explanation of the process."

If you were to compare this next description of the shot-making operation with the one given on November 26, 1856, you will see that the amount of lead that was melted at the top of the tower had increased:

> *At the top of the tower, in a room about twelve by fourteen feet, a small brick and iron furnace is built, upon which is an iron kettle capable of holding about ten or twenty gallons of melted lead, which is then poured into a copper strainer, and being suspended over the opening in the flooring of the tower, the liquid drops through, and down a distance of one hundred and fifty feet, into a large tank of water prepared to receive it.*

The editor then continued with the description of what happened to the new shot pellets once they had cooled in the tub of water at the bottom of the Shot Tower: "By the time the hot lead reaches the tank, it assumes a globular or shot form and on coming in contact with the water, it is instantly cooled and hardened. The shot are then placed in a barrel shaped tank, which, by an appliance of machinery, is made to revolve, thus cleaning or polishing them."

Once they were all shiny, the good shot had to be separated from the bad. Here is how it was done:

> *They* [the shot] *are then rolled over an inclined plane in an oscillating chest with strainer pans for separating and assorting the different sizes. They are then weighed and bagged. We are informed by Mr. Langworthy, that with the present force of eight hands, and other facilities, they manufacture about five thousand pound weight daily.*

(Editor's note: The reader will remember that in a previous chapter, the shot tower in Spring Green, Wisconsin, was described as being capable of producing between six and eight *hundred* pounds of finished shot per day.)

After his shot-making tour was over, the *Daily Northwest* editor had a chance to look out from his location on top of the highest structure in town, just like the reporter for the *Herald* had done earlier, and he commented about how what he saw should be preserved:

> *We were struck with, and really delighted at the beautiful prospect from the top of the tower. The view of the city is certainly one of the best to be had. If there is an artist in our midst desirous of taking a correct, extended and beautiful view of Dubuque, we would advise him to select the top of the shot tower as his point from which to do so.*

Another story about the Shot Tower was published in the *Daily Herald* on April 14, 1859. It is interesting to note that by this time, the tower had yet another pair of operators, Peleg Tallman and William Hyde Clark. The paper reported, "Messrs. Peleg Tallman & Co. have put the idea to practical working, they have leased the Shot Tower at the foot of Seventh street and are now manufacturing a better article of shot than can be bought in St. Louis or New York."

Also interesting to note is that the Shot Tower is now measured as being 130 feet tall. The *Daily Herald* reported:

> *The Tower is 130 feet high. In the lower rooms are the Furnaces, Engines, sorting and polishing machinery, and the "putting up." The lead is melted and one pound of arsenic is added to every hundred pounds of lead. It is drawn up to the top of the tower and poured into a sieve, from which it falls to the first floor in a reservoir of water. From the reservoir it goes on the dryer and thence to a revolving cylinder in which it is polished. It is then taken out and poured on an inclined plane. At the bottom of this is another inclined plane which is separated from the first by a distance of a couple of inches. The pellets that are perfectly spherical will gain momentum enough in passing down the flat inclined plane to carry them over the space which separates it from the one below. If a pellet is a little lame however, it comes down very slowly, aided mayhap by a boy and a brush, and falls between the two inclined planes into a box and is again to be re-melted.*

The article then explained what happened to the good shot after it was separated from the bad and added that perhaps more products than just lead shot would one day be produced:

> *Those that go over both planes are of all sizes. They are then taken out and put into a sort of sifting machine. This looks like a chest of drawers—the top drawer is perforated a certain size, the next below it a smaller size, and so on down. The shot is poured into the top and the whole machine driven backward and forward by machinery. Of course, all the pellets below a certain size corresponding with the holes in the top drawer fall through and so on down—each drawer retaining a certain sized pellet, say "B.B.," "No. 1," etc. They are also making arrangements to manufacture sheet lead and tubing. We are glad to be assured that they are receiving liberal orders.*

There is a lot still to be discovered about the history of the Shot Tower, but much can be learned just by going through old newspapers. However, at certain times reading these articles was difficult to do because some of them were faded or smeared when they were printed. Also, I was surprised how the articles were written, sometimes with words that we don't use much these days. However, as I read these explanations of how the shot-making process took place at the Shot Tower, I could almost see it happening in my mind. I hope you now have a better understanding of who owned the Shot Tower and how shot was made there over 150 years ago.

Chapter 21

In One Hand and Out the Other: The Owners of the Shot Tower

By Spencer David and Carol Cross

The Shot Tower in Dubuque made shot for less than six years, and in some cases, it sat idle because of the Mississippi flooding or because it was sold to other owners. The tower was owned by a number of local businessmen who tried to take advantage of all the lead that was being mined in Dubuque. This chapter focuses on the men who did what they could to turn gray lead into cold cash.

Before the Shot Tower was built in 1856, the area of Dubuque was successful at mining lead, but it did not have a manufacturing industry to make lead products such as shot. The lead that local miners removed from the ground was shipped to St. Louis, where it was made into shot. Citizens of Dubuque actually had to purchase their shot from distributors in New York at a greater cost. This was silly because the lead used to make the shot often came from Dubuque. To solve this problem, a local gunsmith, George W. Rogers, decided to change things. The *Dubuque Daily Express & Herald* ran a story about him on December 3, 1855. It said in part, "A Shot Tower in Dubuque—Mr. G.W. Rogers has been in contemplation for some time to erect a shot tower in the city, and we learned that he has now purchased the ground and is making preparations to build."

Just two months later, on February 13, 1856, Rogers took out a mortgage with the Dubuque Harbor Improvement Company for $10,000 to build the Shot Tower on a man-made island right by the Mississippi

River shoreline in Dubuque. The tower itself cost $7,000, and another $3,000 would be spent for building an additional attached side building that would become a machine shop to house the equipment needed in making shot. It would include a pulley system to raise and lower blocks of lead to the top of the tower where they were melted and pots to pour the molten lead down the shaft.

On August 20, 1856, about six months later, the *Dubuque Daily Republican* reported that George W. Rogers, manufacturer of guns and cutlery, would soon have his tower erected on the island, at the foot of the levee. It would be capable of manufacturing three to five tons of shot each day and be driven by a three-horsepower steam engine. The paper was also very specific in reporting the Shot Tower's dimensions: "It will be built of stone, 25 feet at the base, and ten feet at the top. The tower will be 150 feet in height from the base." The entire project took several months to complete. Rogers finally began regular operations of making shot on December 24, 1856, and sold his product "at St. Louis prices."

On July 1, 1857, the *Daily Republican* reported that the Dubuque Shot Tower was now a large business. Lead shot production was operating on a grand scale, and Rogers exported what he made to the East by using the railroad. He also sold "up-stream" shot to the northern section of the country without the "down-stream" prices like in St. Louis. In other words, his shot was the same quality that was made in St. Louis, but people could choose to buy from him because it was less expensive and there was less distance between them, so they could get their hands on it faster. But soon, the Shot Tower would have a new owner.

By September 22, 1857, newspaper accounts said that "Messrs. Cook and Langworthy" were operating the Shot Tower. The editor of the *Daily Northwest* visited the site, saying he got the approval from tower owner Langworthy to see how shot was made. He described his long climb up the tower using ladders and discussed the shot-making process that was explained to him by Mr. Langworthy, whom he met at the top of the tower. The editor said the drop was 150 feet and that business was very good because Langworthy told him that with "the present force of eight hands and other facilities," they could put out about five thousand pounds of shot every day. However, the Shot Tower would soon change ownership again.

In the spring of 1859, businessmen Peleg Tallman and William Hyde Clark were now reported as making shot at the tower. These men took over the production and improved it by installing new machinery, but their shot-making only lasted three months, even though there was plenty of lead ore in this area. The *Morning Sun* of March 26, 1859, reported the activity:

> *Messrs. Tallman & Clark tested the newly arranged machinery in the shot tower, yesterday afternoon, by putting it in motion and dropping shot. It works to perfect satisfaction and the establishment will commence to manufacture for the market on Monday next. Success attend the enterprise, which we have no doubt will follow with such men at the wheel, and in the native home of the material.*

According to historian Jim Jacobsen, "this is the only account that associates Tallman and Clark as partners in the shot-making business."

Jacobsen also explains that Peter Cook and William Langworthy, of the Dubuque Harbor Improvement Company, were said to have become the owners of the Shot Tower once again when they restarted the shot-making production at the Shot Tower in October 1859. He adds that their operation was a short one because the Shot Tower "was sold to yet another unidentified party and the Dubuque Harbor Improvement Company was forced to foreclose on the new owner in January of 1860."

A month later, in February 1860, the city marshal seized the property. A public auction was held on March 14, 1860, and the Shot Tower was sold to the highest bidder. A man named Gilbert Woodruff of Rockford, Illinois, made the first bid, even though he did not have a real desire to win the bidding. However, his offer was the only one presented, and his bid of $1,653 made him the legal owner of Lot 24, where the Shot Tower stood. The next day, March 15, he turned Lot 24 over to Dubuque businessman Julius K. Graves, who had bid $1,200 for the tower at another auction. By the end of that year, Graves also purchased Lot 22 and Lot 23, which were next to the Shot Tower, for $5,450. The Dubuque Harbor Improvement Company had divided up the land by the river into blocks and lots in the hopes of selling them to businesses. Graves eventually would be known as the last Dubuquer to own and operate the Shot Tower.

According to Jacobsen, there was no news that could be found in the papers about shot production at the Shot Tower in 1860. But if the Shot Tower was not making shot, it wasn't because of a lack of local lead production. The total lead exports in Dubuque in 1860 totaled 67,225 pigs of lead. (A pig is a seventy-pound block of lead and was sold for $4.75 each.) That was the equivalent to nearly $320,000 in 1860 dollars.

MARSHAL'S SALE.

STATE OF IOWA, DUBUQUE CITY, SS.—Notice is hereby given that on the 14th day of March, A. D. 1860, at 10 o'clock A. M., will be sold at public sale to the highest bidder for cash, at the front door of the CourtHouse in the city of Dubuque, the following described property, taken on an special execution issued out of the office of the Clerk of the City Court of the said City of Dubuque, in favor of the Dubuque Harbor Improvement Company and against George W. Rogers et al. to wit:

Lot 24, in block 7, in the Dubuque Harbor Improvement Company's addition to the city of Dubuque, with the improvements thereon.

To satisfy said writ of special execution and all accruing costs. THOS. FLEMING, City Marshal.

COOLEY, BLATCHLEY & ADAMS, Att'ys.

Dubuque, February 13, 1860. d4w(137

A legal notice published in February 1860 announcing the impending sale at auction of the Shot Tower scheduled for March 14, 1860. *Dubuque Express & Herald.*

The Civil War began on April 12, 1861, when Rebel cannons were fired on Fort Sumter, in South Carolina. Here would be an opportunity for anyone who owned the Shot Tower to make a profit by selling ammunition to the Union army. Our class has heard many stories about the Shot Tower making shot for the Civil War. What is the truth? One man seems to have enough information to settle the issue.

Here is the answer to the big question many people ask when talking about and discussing the Shot Tower: did the Shot Tower ever produce shot or bullets for military purposes during the Civil War? In a word, no! Historian Jacobsen did his own investigation, searching through many old documents, including Record Group 101, Adjutant General, Invoices of Ordnance and Ordnance Stores, 1861–1865, Box 1, and he says that even though J.K. Graves made a generous offer to the governor of Iowa just after the Civil War started to turn over the production of shot made in Dubuque to the Union army to fight and defeat the Confederates, there is no evidence of the Dubuque Shot Tower producing any kind of ammunition that was purchased for the Union in the Civil War.

The newest owner of the Shot Tower, J.K. Graves, then announced that he was going to start shot production once again, and it was reported on June 15, 1861, in the *Daily Times*:

> *The original owners* [of the Shot Tower]...*were unable to run it after it was completed. It has been sold and re-sold upon different decrees of the court and on mortgages until at last it is in the hands of our enterprising young townsman J.K. Graves, Esq. He has associated with him other men of experience and these uniting with him will start again in the business of making shot of all sizes, on Monday next. P. Tallman, Esq.,* [the Shot Tower's] *former superintendent, will have charge of the manufacture.*

Peleg Tallman was picked by Graves to be in charge of manufacturing. Tallman was once the owner of the Shot Tower with William Hyde Clark. Even though there is no evidence that shot from the tower was used in the Civil War, it seems that he and Graves wanted to come up with new ways to increase sales and get more customers. To make their product more popular and easier to buy, Tallman decided to change the way shot was produced for market. The July 3 edition of the *Daily Herald* noted the following:

> *Heretofore it* [shot] *has always been put up in twenty-five pound sacks, but he* [Tallman] *will put up half sacks, containing twelve pounds, and quarter sacks with six pounds. These small packages will be very popular with sportsmen, as many gentlemen who go out for a day's shooting, do not wish to purchase a full sack.*

Jim Jacobsen says in a bit of a humorous way that it is interesting to note that "sportsmen" and "gentlemen" are mentioned as likely purchasers of shot, not Union soldiers or anyone else in the Union army.

As of January 1862, the Shot Tower was a very busy place and was making a lot of shot. However, the cold winter weather made it impossible for anyone to ship products up or down the frozen Mississippi River, and this included the bagged shot that was coming from Dubuque. The following accident account in the newspaper proves that the Shot Tower was in full operation at the time. This was reported on January 9, 1862, in the *Daily Times*:

> *A serious accident occurred at the Shot Tower Tuesday afternoon, at about three o'clock, occasioned by the floor of the finishing room attached to, but not a part of the Tower. Owing to the impossibility of shipping across the river for the past few days the quantity of shot has accumulated together with the large quantities of lead stored on the premises, caused the floor to give way, precipitating three of the workmen together with about a thousand bags of*

> *shot, eighteen tons of lead pigs and cakes, two cauldrons of molten lead, weighing about fourteen hundred pounds, as also miscellaneous articles used in the room, including one large stove, and a large furnace, under full blast.*

The newspaper wanted its readers to know that this accident was not caused by carelessness by anyone who worked at the tower but was proof that production was ongoing and would resume soon:

> *The blame rests of course in storing such quantities of lead in the building, the workmen were, however, governed by their own ideas, as to whether it should be stored there or left piled outside the building, so that no fault can be attributed to anyone. The loss will be small—aside from the delay—as the lead shot can be re-melted.*

The article went on to say that the damage to the Shot Tower's outbuilding "will be remedied at once so that business will proceed as usual after the present week."

By the spring of 1862, the Shot Tower was in full production once again, and it was reported in the *Daily Times* on March 19. The Graves Tower was producing 100 to 150 sacks of shot per day. The paper also said that Graves was going to install new machinery, hoping to keep up with shot demands. For example, a Chicago business wanted a lot of shot, and other companies were also standing in line. The paper said:

A vignette of the Shot Tower and its outbuilding that complemented a larger engraving of Dubuque's waterfront, published by W.J. Gilbert in 1858. Note the figures approaching the inclined walkway to the outbuilding's front door. *Dubuque County Historical Society.*

> *This company's works are in full blast at the tower, and they are turning out from 100 to 150 sacks per day. Mr. Graves is introducing improved machinery and otherwise adding to the advantage of this factory, hoping thus to be able to keep up to his orders. One firm in Chicago, agreed to take one thousand sacks per week during the whole of 1862. Burbank and Company of Saint Paul, Minnesota, as soon as the river navigation opens wish to be supplied with several hundred sacks weekly for the trade of the Red River of the North. These are only two of our great customers of our shot tower...Other orders from a distance are pouring in more rapidly than they can possibly be filled.*

Anyone reading these newspaper articles up to this point would assume that even though there were occasional floods and accidents, things seemed to be going very well for businessman J.K. Graves at the Shot Tower. However, on July 18, 1862, the *Daily Times* shocked its readers with the following story:

> *THE SHOT TOWER IS SOLD—J.K. Graves, Esq., informs us that the Shot Tower in this City has been sold to Mr. Chadbourne of the Saint Louis Shot Manufacturing Company. It has been purchased because it interfered so much with their trade, and we learn with so much regret it is their intention to take out the machinery, board up the doors, and close the establishment merely because it has become too powerful a rival. This directly affects our prosperity to a certain degree, and it rests only with our capitalists to say whether the manufacture of shot shall be abandoned entirely by the citizens of Dubuque.*

Something had to cause Mr. Graves to decide to close down his Shot Tower. The St. Louis purchase price of $3,000 was hardly an offer one couldn't refuse. Graves had much more money than that tied up in the Shot Tower, along with the additional property and the improvements he made on it. He also had plenty of orders to make shot. Why did he agree to close his doors and sell?

Now that Chadbourne & Forster had bought the Shot Tower, the building that had gone through the hands of George W. Rogers, Peleg Tallman and William Hyde Clark, Peter Cook and James Langworthy, Gilbert Woodruff and finally J.K. Graves, would soon be abandoned for good. Would this be the end of shot-making in Dubuque? Did all these shot-making businessmen really give up without a fight? Who would later come to own the tower, and what would it be used for?

Chapter 22

The Life and Times of J.K. Graves

By Sean Bosveld and Carol Cross

Julius K. Graves was born on September 29, 1837, in Keene, New Hampshire. He went to school in his hometown but left at the age of seventeen. He was heading west to find work when, in 1855, he came along a town called Dubuque, Iowa. He soon found a job at the M. Mobley & Company Bank as a cashier; he was then only eighteen years old. Just three years later, the bank reorganized in 1858 and then merged with the State Bank of Iowa. Dubuque was its first branch, and Graves served as the vice-president and general manager. Graves also served on the State Bank's Board of Control as a director. In later years, Graves was a major organizer of both the National State Bank and Commercial National Bank. However, he had other interests as well.

At the age of twenty-two, J.K. Graves also became an owner and operator of the Dubuque Shot Tower in 1860. He bought it at auction from a Rockford, Illinois businessman named Gilbert Woodruff for $1,200. What is interesting is that Woodruff had bought the structure at the local marshal's sale the day before, when he paid $1,653 for it. Eight months later, Graves bought up two additional lots next to the Shot Tower site (Lot 22 and Lot 23) for $5,450. These empty lots were actually more expensive than the lot where the Shot Tower stood. So who was J.K. Graves, and what kind of a man was he?

Julius Graves was someone people in Dubuque looked up to as a serious and honest businessman. Harry C. Slichter, a reporter for the *Dubuque*

Telegraph Herald, wrote an article about him in 1963 and explained how he became the owner of the Shot Tower:

> *He bought the Shot Tower almost by accident. In newspaper accounts, he tells how he was passing by an auction sale (it happening to be the Tower) and started the bidding at $1,200. His was the only bid, and he unexpectedly had a shot tower on his hands. (An "elephant," he called it). But he put it to work, having a plentiful source of raw material from the Dubuque lead mines.*

Even though Graves was new to the business world of shot-making, he was a very smart man and could figure things out quite fast. He and his employees were soon producing a good amount of shot and were competing with Chadbourne & Forster, a shot-making company in St. Louis, Missouri. Oddly enough, Chadbourne & Forster was Dubuque's source of shot before the Shot Tower was built.

On April 14, 1859, the *Dubuque Daily Herald* published an article that explained how the Shot Tower could save local citizens a lot of money if a business was created that could compete with another shot-making company in St. Louis. It asked a question: why should customers in Dubuque have to purchase lead shot from New York that originally came from Dubuque? The paper said, "The *Daily Herald* at various times has urged the necessity and economy of lead manufactures in the city upon the plea that it is folly to send our lead to St. Louis, then to New York, at which place it was bought by our dealers. This enormous addition of freight was just exactly so much money wasted."

Peleg Tallman became the man who would be in charge of making shot in Graves's tower. For the convenience of sportsmen, he decided to improve the way shot was packaged. Instead of the usual twenty-five-pound sacks, he created half sacks containing twelve pounds and quarter sacks weighing six pounds. These smaller bags became very popular with sportsmen because they didn't have to spend extra money to purchase a full sack of shot and were easier to carry when hunting. It seems that these men agreed with the *Daily Herald* that a shot-making business would be good for Dubuque.

During the year 1860, there was nothing written about shot being made by Graves at the tower. However, according to historian Jacobsen, this does not mean there was no activity; it is just that there were no news articles about the tower or what was going on there. In the middle of all this, Graves got married to Lucy C. Robinson from Salem, Massachusetts. They eventually had five children: Luis K., Julius K. Jr., Sybil, Sara and Marjorie.

On April 12, 1861, Confederate soldiers fired on the Federal post called Fort Sumter, located in the bay off the shore of Charleston, South Carolina, and started the American Civil War. Would there be a chance for Graves to make money by selling shot to the Union army to defeat the Rebels? After the war began, the first mention of the Shot Tower in 1861 came on May 24, when the *Daily Times* reported that high water on the Mississippi River was preventing Graves from producing shot at the tower but that as soon as the water level went down, production would begin again. The paper also took the opportunity to explain that shot made locally would be cheaper to purchase and that it would be foolish to buy it any other way:

> *Shot Tower—We are informed by one of the proprietors of this fine building that as soon as the high water subsides in the river, they will commence the manufacture of shot once more. It is our sincere hope that as long as shot are being made anywhere, this institution will continue to run to its full capacity. It is entirely wrong to pay half a dozen freights and profits on the products of our lead mines before it gets to our hands ready for consumption, when it might be made up here at our doors as well.*

In the same article, the *Daily Times* reported that Graves was going to use his Shot Tower to display his patriotism: "J.K. Graves, Esq., a gentleman who is largely interested in this enterprise, purchased a large flag the other day which cost him $40, which he intends to raise upon a flag pole from the top of the shot tower on the receipt of the news of the first battle and victory for the Union. Good for him."

Unfortunately, the Union army did not win many early battles against the Confederates, and it got to the point where J.K. Graves decided to not wait any longer. He went ahead and displayed his Star-Spangled Banner on May 30. The *Daily Herald* reported the story the next day:

> *The Highest Flag in Iowa—The patriotic proprietors of the Shot Tower raised yesterday a fine S.S.B.* [Star-Spangled Banner], *on a staff above the top of the Tower, 160 feet from the ground. We doubt whether there is a pole of that height in the State. At least, Dubuque will claim the honor of the highest flag until further news.*

Also in May 1861, Graves sent a letter to Iowa's governor, Samuel J. Kirkwood, explaining that he was ready to put the entire production of his Shot Tower into the hands of the Union army, if he was asked. He wrote:

> *As the sole owner of the "Dubuque Shot Tower," I take this occasion to tender the use of said property to the Government for the manufacture of such supplies of Bullets and other Leaden missiles of war as may be required during the continuance of the present troubles—without money and without price. The Tower is in good running condition and is at your immediate service.*

According to Jim Jacobsen, the governor did not respond to Graves's letter. This is a strong indication that shot from the Shot Tower did not get sold to the Union army to use in the Civil War. If Graves had gotten a contract, the newspapers surely would have reported it because that would have been major news. One of Kirkwood's aides wrote a brief note to the governor explaining to him just who J.K. Graves was: "Mr. Graves has just suggested

Samuel J. Kirkwood (1813–1894). Iowa governor from 1860 to 1864, Kirkwood became nationally known for procuring soldiers and supplies from Iowa for the Union army. He later served as secretary of the interior in the Garfield administration. *Library of Congress.*

the idea to me of offering the use of his 'Shot Tower' &c. to you for the patriotic purpose of giving them 'a little more grape.' He is one of our best businessmen & has every facility to serve the people in this emergency... whatever he promises will be done." But Kirkwood never replied to Graves.

Historian Jacobsen also reported that to help with the war effort, the Graves brothers (J.K. and R.E. Graves) loaned the State of Iowa "considerable sums necessary to raise and equip military volunteers from the Dubuque area." Graves himself "took the lead" in raising, equipping and paying for the military establishment that was soon called Camp Union. But there is no evidence that Graves won any kind of contract to make shot for the Civil War.

On October 24, 1861, the *Daily Herald* sent a reporter to Graves's Shot Tower, and he described in detail the shot-making process. This time, there was a lot of attention paid by the reporter to one of the main ingredients of making shot—arsenic. The reporter must have survived his contact with this deadly material, because his story ran the next day:

> *We visited the Dubuque Shot Tower, J.K. Graves proprietor, on yesterday afternoon, where we witnessed the novel sight, to us at least, of making shot: the first room below the one we entered contains the finishing department. We observed in this room an immense quantity of shot in various stages of perfection, from that in which it is taken out of the water tank or reservoir, to that in which it is left after having received its bright polish, and being ready for market.*

The reporter then described the different kinds of shot lying around all over the place. It is interesting to note that the reporter thought that Graves used arsenic to coat the shot. We have learned that arsenic was added to the molten lead to make it brittle, which made it easier to separate and then form into pellets as it fell through the holes in the sieve and then down the length of the Shot Tower. Unless the reporter was exaggerating, the amount of arsenic that was sitting in the tower must have been impressive because he reported that there was enough of it there to kill over ten thousand men:

> *We saw shot on the floor, in boxes, tubs, shelves, drawers and bags, and of all sizes, from that of a very small pin head to that of an ordinary pill sugarcoated. Mr. Graves coats his shot with a material of a somewhat different character, namely* arsenic. *We saw a box in one corner of the room holding enough of this "pleasant material" to "poison a dozen regiments of*

> *soldiers* [a regiment was made up of about one thousand men]." *After examining this room to our satisfaction, we were next shown up into the tower* [quotes and emphasis in original].

Then the reporter described how he got to the top of the Shot Tower and how Graves got the metal up there to drop it after he blended in some of that dreaded stuff, arsenic:

> *After going up some few hundred steps of stairs, our friend J.K. led us into the casting room. The lead, after having been thoroughly mixed up with that deadly stuff,* arsenic, *is by means of a dumb waiter, hauled up to an elevation of about 170 feet from the ground into this room, where it is melted and poured through sieves of different sized apertures down through the distance of about 150 feet into a reservoir below* [emphasis in original].

The reporter then talked about what the liquid metal looked like as it fell through the Shot Tower's shaft:

> *We watched this operation with much astonishment. The lead falls through this sieve in as many beautiful and steady streams as there are holes in the sieve. At the distance of about 50 feet, those streams begin to break up in small detached pieces, and gradually become round as they fall, so that by the time they reach the water, they are separated into innumerable bright small globules of shot. We descended and watched the process of sorting the various sizes and bagging them up ready for the purposes for which they are made.*

At this point in the article, the reporter began to sound more like an advertisement for the Shot Tower and the products made there: "Notwithstanding the fact that they make over 2,500 lbs. per day, the demand is so great that they can scarcely supply it. We regard this as one of the best paying establishments of any kind in Dubuque."

In the years that Graves owned the Shot Tower, what he called the Dubuque Shot Manufacturing Company was capable of producing about five thousand to eight thousand pounds of shot daily. Usually, production amounted to 100 to 150 sacks per day. A March 19, 1862 article in the *Dubuque Daily Times* reported that Graves had many customers and that they were buying everything he could make:

> *Dubuque Shot Manufacturing Company—This company's works are in full blast at the tower, and they are turning out from 100 to 150 sacks per day. Mr. Graves is introducing improved machinery and otherwise adding to the advantage of this factory, hoping thus to be able to keep up to his orders. One firm in Chicago agreed to take one thousand sacks per week during the whole of 1862. Burbank & Co. of Saint Paul, as soon as the river navigation opens, wish to be supplied with several hundred sacks weekly for the trade of the Red River of the North. These are only two of the great customers of our Shot Tower. No small amount is sold by the local agents, Smith & Cannon, and other orders from a distance are pouring in more rapidly than they can possibly be filled.*

With all that had happened so far with shot production in Dubuque, nothing was said yet about a competing St. Louis shot-making company, the primary reason why the Shot Tower was built in Dubuque in the first place. Historian Jim Jacobsen reports that the last local newspaper reference to the Shot Tower and its production of shot dates to mid-July 1862. He says the content of that story suggested that the Shot Tower might be closed down for some reason, making it "an abandoned facility that was vulnerable to the visitation of local youths." An article on July 13, 1862, from the *Daily Times* said:

> *THOU SHALL NOT STEAL—A boy stole a chunk of arsenic from the Shot Tower a day or two ago and took it home. There he threw it on the ground in the garden and went to dinner. Soon after he had finished his meal he went out doors and was dismayed to see sixteen of his father's Shanghais* [a breed of chicken] *dead upon the ground, and as many more on the point of giving up the ghost. Served him right.*

Jacobsen thinks that in order for anyone to be able to steal a sizable chunk of arsenic from the Shot Tower like this youngster did, it must have been shut down and no one was there to guard the property. This turned out to be the last article about shot production at Graves's Shot Tower. Despite the large amounts of shot he was producing, J.K. Graves sold the Shot Tower to a St. Louis shot-making company. The *Daily Times* of July 18, 1862, reported the following news to the citizens of Dubuque:

> *THE SHOT TOWER IS SOLD—J.K. Graves, Esq. informs us that the shot tower in this city has been sold to Mr. Chadbourne of the Saint Louis*

> *Shot Manufacturing Company. It has been purchased by them because it interfered so much with their trade, and we learn with much regret it is their intention to take out the machinery, board up the doors and close the establishment because it had become too powerful of a rival.*

Chadbourne & Forster paid Graves $3,000 for the building. After selling the Shot Tower to this St. Louis company, Graves signed an agreement that he would not use the tower to make shot or build another one for ten years. Chadbourne & Forster bought the Shot Tower to shut it down and have a competitor out of the way. Although they promised to spend $5,000 on it and get it back to work producing shot, they boarded up the windows, took out the machinery and told their agents to lock the door and throw the key in the river.

However, it seems that Graves still had plans to make shot, only this time he would not need a tower but something else that would give him the height he needed to drop molten lead. The *Daily Times* picked up the story:

> *No sooner was the shot tower sold than a party of gentlemen took a wagon load of all necessary implements and started out to make a trial of an idea that had been whispered about on the sly in the city for some months past. They had melting pots, lead, ladles, ropes, tubs, firewood, etc., etc. Going out in the vicinity of General Booth's residence, they built a fire within a short distance of an old mining shaft, and proceeded to melt a quantity of tempered lead. The shaft was some 140 feet in depth. Attaching ropes to a washtub partly filled with water; they let it down to the bottom of the pit. This was for the purpose of catching the pellets of lead as they dropped down. A sieve was adjusted as nearly over the tub as possible, and when the lead was perfectly melted, the experiment was completed by pouring it through in the same manner that is pursued at the shot tower.*

The tub was then slowly brought to the surface and its contents were taken out. The shot made by this method using "shot holes," as the shafts were called, was of equal or better quality than the shot made at the Shot Tower. This was accomplished with a very small amount of cash, usually about $500, instead of the $15,000 necessary to build a tower.

Historian Jim Jacobsen revealed that two other successful shot-making companies were created as a result of this mine shaft experiment. One of them included J.K. Graves. He was joined with William Carter and General

Caleb H. Booth. The August 3, 1862 edition of the *Daily Times* reported on this new shot-making operation. Graves was back in business less than a month after he sold the Shot Tower. It appeared he found a 130-foot shaft and only had to build a 20-foot extension over it to reach the 150-foot height needed to make shot:

> *THE NEW SHOT TOWER—Messrs. Booth, Carter, and J.K. Graves commence the erection to-morrow of a new two story brick building over the shaft just west of Gen. Booth's residence, to be devoted to the manufacture of shot. This company is a strong one, and composed of gentlemen who will not allow any small sum to buy them off from the track. This project has been rapidly conceived, and will be just as rapidly carried forward to completion, and once more will Dubuque supply the great Northwest with shot.*

Julius Graves did not spend all of his time making shot and counting money, however. By the age of twenty-five, he was one of the most respected businessmen in the city of Dubuque. He was also a very busy man after the Civil War ended in 1865. In 1866, he was elected mayor of Dubuque. According to local historian Bill Hammel, back in those days, mayors in Dubuque served for one year and then could choose to run for reelection if they wanted. Hammel also says that it was customary for the fire department to name new engines after the mayors. In recognition of having the first steam fire engine used in Iowa, the "J.K. Graves," named after him, Graves donated his official salary as mayor to the fire department. This noted businessman also founded the Key City Gas Works, which was how the city of Dubuque was first lighted, and became its president and chief stockholder. He also took a very active part in the organization and construction of the Dubuque Street Railway and served as its president in 1868.

In 1876, J.K. Graves was elected to serve in the Iowa State House of Representatives. In 1881, he was elected to serve four years as a state senator. He also became involved with railroad construction in 1870, serving as president of the Chicago, Clinton & Dubuque Railway Company; the Chicago, Dubuque & Minnesota Railroad; and the Iowa Pacific Railroad. During the time he was in this business, he supervised the construction of the railroad that ran from Dubuque to Chicago and one from Dubuque to Minnesota. This was a great method of transportation for the products these cities needed, like coal, shot and other items, in order to grow. Graves also served as president of the Iowa Pacific Railroad Company. He traveled throughout the United States, including a trip across the Plains to New Mexico, as a special commissioner appointed by the Department of the Interior to hold council with six thousand Navajo Indians. He also visited Europe twice.

Being a banker can be a real challenge, especially if you have to walk back and forth from your office downtown to your home on the bluffs in Dubuque for lunch. Mr. Graves figured it cost him over an hour of time every day and came up with an idea. In 1882, he designed and built a private elevator to go up and down the bluffs. At the time it was constructed it was called the Fenelon Place Elevator, but we know it today as the Fourth Street Elevator.

The original Fenelon Place Elevator, built by J.K. Graves in 1882 to reduce his daily hour-long walk from downtown to his hilltop home for lunch, is today the nation's shortest and steepest stretch of active railroad. *Center for Dubuque History.*

It is recognized for being the shortest and steepest stretch of railroad track in the United States—only 286 feet long, with an incline of nearly seventy-four degrees. Even though the elevator was destroyed by fire in 1884, Graves rebuilt it and opened it up to his neighbors, who paid a five-cent fare to use it. The elevator caught on fire again in 1893, but this time, citizens created the Fenelon Elevator Company, bought the system from Graves and had it restored. You can still use it today.

When J.K. Graves died in 1898, many people in Dubuque stopped to remember what his life meant to our city. The December 10, 1898 edition of the *Dubuque Daily Times* gave its readers a brief history lesson as it reviewed Graves's life in his obituary. It said: "In the early days of the history of the city, he was looked up to by the younger generation as a splendid type of the self-made man, and enjoyed the confidence and esteem of the staid businessmen of the city."

Beginning with Graves's childhood, the paper recalled how he got introduced to the world of business and banking: "He came to Dubuque in 1855 and secured the position of cashier in the banking house of M. Mobley & Co. He was then eighteen, and at the age of twenty-five he had ranked as one of the foremost businessmen of the city." Most people think Graves was only responsible for the building of the Fourth Street Elevator and owning the Shot Tower, but he did so much more for the city of Dubuque and its citizens, as the *Daily Times* pointed out:

> *Mr. Graves has been prominent in the organization of great many of the public institutions that have given Dubuque prominence. In 1868 he took an active part in the organization of the Dubuque Street Railway Company and was its president. He started Key City Gas Company in 1859, and in 1866 built the work that supplied the city with lights. Mr. Graves was one of Dubuque's pioneer railroad builders, and it was largely due to his push and enterprise that this city secured the splendid railroad facilities it has.*

Even though he became wealthy, J.K Graves did not turn his back on anyone who needed help. The paper said this about him:

> *Those who knew J.K. Graves best assert he was the soul of honor in his dealings. He was all business in his business dealing and was quick to see and improve an advantage, but in his private life many a needy family revered him and his estimable wife, for it is a frequent saying that when J.K.*

Julius K. Graves (1837–1898). Respected local businessman and philanthropist, last owner/operator of the Shot Tower, former mayor of Dubuque, builder of the Fenelon Place Elevator and great-great-grandfather of Central Shot Tower Investigator Sean Bosveld. *Dubuque County Historical Society.*

> *Graves had means no man applied to him in vain, and it is also said of him that whatever faults he had, they were of the head, not the heart, and that had he been less tender hearted and less ready to help a friend his business career would have ended differently.*

In many ways, a person can be best evaluated after he dies because no one will be afraid of him any more. Also, over time, a person's true character is appreciated more and more. After J.K. Graves passed away, the *Daily Times* reported his life's journey, which already has been referred to in this research. However, I believe the very first sentence of his obituary is the greatest tribute and a fitting way to end this chapter: "J.K. Graves is dead, but the memory of his past life and works, his deeds of philanthropy and the life and vigor he infused into the commerce of the city will ever be one of the brightest pages in the history of the lives in Iowa's great men."

J.K. Graves was an exceptional man. He was one of the few people in town who could say they owned and operated the Shot Tower. He was also very respected and appreciated by the citizens of Dubuque, and I am happy to have had the chance to write this chapter about him. This research means a lot to me and my family because Julius K. Graves was my great-great-grandfather.

Chapter 23

The Shot Tower Gets Shut Down

By Tim Kenniker

What happened to cause the Shot Tower to be shut down in 1862? It is commonly blamed on Chadbourne & Forster, a rival shot-making company in St. Louis that was trying to wipe out any competitors. We know the St. Louis Shot Company purchased the Dubuque Shot Tower in 1862 because it "interfered too much" with its trade. The location of the Dubuque Shot Tower and its ability to undercut prices of the St. Louis competition played a major role in the Dubuque Shot Tower being bought by Missouri businessmen. Chadbourne & Forster's intentions were to pull out the machinery and board up the doors "merely because it had become too powerful a rival."

The accounts of the sale, and the motivation for it, are found in the *Dubuque Daily Times* and the *Dubuque Express & Herald.* However, at the time J.K. Graves sold the tower to his Missouri competition, demand for Dubuque-made shot was overflowing. If orders were pouring in faster than they could possibly be filled, why would he sell the tower? It doesn't seem to make much sense.

Here is some information that shows just how good business was for Graves and the Shot Tower. Early in January 1862, when the weather was cold and the Mississippi River was frozen, the Shot Tower's employees were still making and storing shot for future sales. In fact, so much shot was stored in the outbuilding next to the tower that an accident occurred, leaving several men shaken up and one seriously hurt. The floor of the finishing room actually collapsed under all the weight of the bags of lead shot, lead pigs and other materials and tools used in shot production. Three men were working at the time of the accident, but perhaps by a miracle, only one of them wound up seriously hurt. The *Daily Times* of January 9 described the details of the injuries he suffered and who treated him:

> *Strange to say, amidst the ruins thus created, but one of the workmen—Mr. Richard Dillon—was injured. He was found with his leg broken and his head somewhat bruised, surrounded by huge piles of lead and a large beam across his neck, so that his escape from instant death was truly miraculous. He was promptly cared for by Drs. Lay and Lewis, and is doing well.*

Graves's shot-making effort was very profitable, at least through March 1862. He may have seen his opportunity to expand his business even more with the Civil War that had begun in April 1861. However, the only reference to any possibilities of war-related production was a letter sent by Graves in May 1861, eight months before the January 1862 accident just mentioned, to Iowa's governor, Samuel J. Kirkwood, with a generous deal concerning the Shot Tower. He offered to make all the shot he could produce available for the war effort "without money and without price." He also said that "the Tower is in good running condition and is at your immediate service." Graves explained that if the government wanted ammunition, he and his workmen were ready to produce all the shot the army wanted because he didn't have to go far to get the lead:

> *Or if preferred I shall be most happy to furnish such war materials delivered at such points, as you may deem expedient and at actual cost. As you are well aware this section of Iowa is the seat or "foundation head" of the lead region. The supply is inexhaustible, which in fact coupled with the facilities of the Tower for the manufacturing of bullets, shot, etc. I commend to your consideration.*

Graves went on to say that now was the time for all good Americans to step up and lend a hand to achieve victory in the Civil War against the Southern rebels. He was offering to supply shot not because he could make a good profit but because he believed very strongly that the rebellion must be squashed:

> *In the present struggle of liberty with barbarism it behooves all that love their country to sink partisan in the Patriot to yield personal emolument* [profit] *and gain to the glorious cause of our flag. This is the motive, which activates me, and I am ready to add my might in advancement of the "common cause" and will gladly furnish that which shall deal death and destruction to the traitorous Southern Rebels "until the last armed foe expires." Your obedient Svt.* [sic] *J.K. Graves* [quotations in original].

Graves almost begged to be allowed to give the Union army unlimited amounts of shot to help win the Civil War. However, there is no documented response from Governor Kirkwood to Graves regarding his generous promised production of shot. This is evidence that shot made at the Shot Tower in Dubuque was never used in the Civil War because it wasn't ordered by anyone in the military, although they could have bought it for a very good price.

Even though J.K. Graves did not get a contract to make shot for the Civil War, almost a year later there was planning in the works to substantially increase shot-making since business was going great. Because Graves was expanding production to keep up with orders, almost anyone could figure out that the selling of the tower would be out of the question for him. The *Daily Times* on March 19, 1862, reported that production at the Shot Tower was running at 100 percent and that Graves was getting ready to make even more:

> *The Company designs erecting this summer a smelting furnace in conjunction with the tower, and be enabled to save a great amount of expense in melting, etc. The original intention of the company was to eventually build machinery and shops for the manufacture of bar lead, lead pipe, sheet lead, and other articles made from lead, but are so overrun with business at present that they find it expedient to devote their whole energies to the manufacture of shot and bar lead alone. It is hoped, however, that the time may come when every pound of the products of our mines may be manufactured here at home.*

The Shot Tower operated by J.K. Graves was doing great business for the next four months until it was reported in the *Daily Times* that it had been bought by Mr. Chadbourne of the St. Louis Shot Manufacturing Company to get rid of the competition. The paper was concerned that losing this profitable business would harm the people of Dubuque. The July 18, 1862 edition said:

> *THE SHOT TOWER IS SOLD—J.K. Graves, Esq., informs us that the shot tower in this city has been sold to Mr. Chadbourne of the Saint Louis Shot Manufacturing Company. It has been purchased by them because it interfered so much with their trade, and we learn with much regret it is their intentions to take out the machinery, board up the doors, and close the establishment merely because it had become too powerful a rival. This directly affects our prosperity to a certain degree, and it rests only with our capitalists to say whether the manufactory of shot shall be abandoned entirely by the citizens of Dubuque.*

In the *Express & Herald*, one week later, an article appeared denying the sale of the Shot Tower to that St. Louis company, claiming a company named Laflin, Smith & Company, which already had a branch office in the area, had purchased the Dubuque Shot Tower for $3,000. I found out that Laflin, Smith & Company was in fact a branch of the St. Louis Shot Company. However, I later learned that this article was incorrect. The editor of the *Express & Herald*, Dennis Mahony, had recently been arrested on charges of disloyalty to the Union cause in the Civil War, so the paper might have been distracted and made a mistake. On July 26, 1862, the *Daily Times* reported, "The Shot Tower passed into the possession of Messrs. Chadbourne & Forster of St. Louis yesterday, and not to Messrs. Laflin, Smith and Co., as the *Herald* stated."

The article agreed that the price paid for the Shot Tower was $3,000 and went on to say that the new owners, Chadbourne & Forster, promised that they would spend $5,000 on the tower and "get it running immediately." However, this didn't happen. Their real intentions were to close it down. Regardless of the sale, the advertising for Graves's "Dubuque Shot Manufacturing Company" continued to run in the *Daily Times* as late as July 27, 1862. The advertisement offered to fill "orders for the various sizes of shot, fitted with dispatch." This leads to the following questions: Why did Graves sell the Dubuque Shot Tower for a measly $3,000? Was running this newspaper advertisement just a waste of money, or was Graves already planning to produce shot in other ways?

Chapter 24

How St. Louis Got "the Shaft"

By Cheryl Leib

It is hard to believe that the famous Shot Tower was in operation for only six short years, from 1856 to 1862. The Civil War was getting more intense by the summer of 1861, and the demand for ammunition seemed to be endless. In addition, the market for animal hunting also put pressure on local shot makers to keep their products coming to meet the growing demand. But something happened in the summer of 1862 that would change the history of the Shot Tower and Dubuque forever.

According to historian Jim Jacobsen, the last reference to the Shot Tower in Dubuque producing lead shot was published on July 13, 1862, when the *Daily Times* ran a story about a theft that had taken place at the tower. He believes that this incident suggests that by the summer of 1862, the Shot Tower had been abandoned for some reason. The article was entitled "Thou Shalt Not Steal":

> *A boy stole a chunk of arsenic from the Shot Tower a day or two ago and took it home. There he threw it on the ground in the garden and went to dinner. Soon after he had finished his meal he went out doors and was dismayed to see sixteen of his father's Shanghais* [a breed of chicken] *dead upon the ground, and as many more on the point of giving up the ghost. Served him right.*

How could the boy have gotten such a large piece of arsenic out of the tower unless no one was working there or guarding it after work? After all, arsenic is poisonous, and that boy also could have walked off with a lot more stuff at the Shot Tower if there was no one to stop him from doing it. What was going on here?

Jacobsen reported that the Shot Tower was sold on July 14, 1862, just one day after this newspaper article about the arsenic theft was published, and that news of the sale leaked out a few days later when the *Daily Times* reported that Shot Tower owner J.K. Graves indeed had sold the structure to a competing St. Louis shot-making company. Had he simply been bought out to eliminate competition? The paper reported:

> *J.K. Graves, Esq., informs us that the Shot Tower in this city has been sold to Mr. Chadbourne of the Saint Louis Shot Manufacturing Company. It has been purchased by them because it interfered so much with their trade, and we learn with much regret it is their intention to take out the machinery, board up the doors and close the establishment merely because it had become too powerful a rival.*

Why was Graves agreeing to sell the building that made shot-making possible in the first place? We can get one idea by reading a small article that was published in the *Dubuque Daily Herald* on July 27, 1862:

> *What is the use of shot towers, when just as good shot can be made in mineral shafts. We were shown some samples of shot yesterday that was made by pouring molten lead down a deep shaft, the shot being equal to any made in a tower. Won't somebody buy up all the mineral holes to prevent shot from being made here? Who bids for the lot?*

In other words, local businessmen had found another way to make shot without the need of a tall structure like the Shot Tower, so they sold it. The *Daily Herald* published a long article on the sale of the Dubuque Shot Tower, but it sounded more like an editorial instead of just a news story. The opinions of the paper were there for all to see, and they were very sarcastic. The article was titled "A Hundred Shot Towers":

> *Messrs. Chadbourne & Forster of St. Louis, paid $3,000 for the Dubuque Shot Tower, for the purpose of having a troublesome rival out of the way.*

> *Although they promised to expend $5,000 on it and set it to work, in order to pacify many dissatisfied businessmen in the city, they of course do not intend to do any such thing. No, they privately inform their agents that they intend to board up the windows, take out the machinery, lock the door and throw the key in the river. Rather than have another sack of shot made in Dubuque, said they, we will put a keg of powder under the tower and blow it higher than Gilderoy's kite.*

We don't know who Gilderoy was or what the paper meant by referring to him flying a kite. But the next part of the article got to the real reason why the Shot Tower was shut down and then bought by the men from St. Louis:

> *The next day after the purchase was completed and the property transferred, the wholesale price of shot, throughout the entire Northwest, advanced 25 cents per sack. Chadbourne & Forster manufacture a thousand sacks per day. The recent advance in price put $250 per day in their pockets, and in twelve days they made enough to pay for the Dubuque Shot Tower. So they can afford to blow it up and not feel the loss.*

After saying that it was not angry with Chadbourne & Forster because they should be allowed to make as much money as they could, the newspaper still didn't like what was going to happen to some local workers—mainly that they would lose their jobs:

> *Now, we have no feeling against C&F. They have, of course, a perfect right to buy up their rivals in business and make as much money as they can. Doubtless they have families to support and they have to look to the main chances to keep the mouths all fed. But this much we object to; we don't like to see a manufacturing establishment that was at work using up the rough products of our mines, blotted out of existence—it materially affects our prosperity.*

Then the *Daily Herald* explained to its readers what was going on behind the scenes in Dubuque just before news that the Shot Tower was sold. It looked as if a group of local businessmen had already found a better way to make shot, and they weren't going to lose a lot of sleep if their precious Shot Tower was to become someone else's property:

No sooner was the shot tower sold than a party of gentlemen took a wagon load of all necessary implements and started out to make a trial of an idea that had been whispered about on the sly in the city for some months past. They had melting pots, lead, ladles, ropes, tubs, firewood, etc., etc. Going out in the vicinity of General Booth's residence, they built a fire within a short distance of an old mining shaft, and proceeded to melt a quantity of tempered lead. The shaft was some 140 feet in depth. Attaching ropes to a washtub partly filled with water, they let it down to the bottom of the pit. This was for the purpose of catching the pellets of lead they dropped down. A sieve was adjusted as nearly over the tub as possible, and when the lead was perfectly melted, the experiment was completed by pouring it through in the same manner that is pursued at the shot tower.

The shot makers in Dubuque were actually trying to make shot without using a tower! It looked as if their experiment was going to work very well, and the paper had the proof:

The tub was slowly and laboriously drawn to the surface, and its contents were subjected to scrutiny of the excited party. "Hurrah! It's a success!" broke forth from lips of one. "By jingo, ain't they bully?" said another. Yes, dear reader, the experiment was a success, and at the present writing a pound or more of the identical shot, contained in a glass bottle, is on the table before us. The shot averages considerable better than that made at the tower. For a few hundred dollars any of these abandoned shafts can be transformed into the best of shot towers or shot holes, which ever you would call them.

Here is where the newspaper got really sarcastic. But after all, who could blame the people of Dubuque for being angry?

By erecting a cheap wooden tower over some of the deepest shafts, shot of the largest size could be made. That the business is profitable is proved by the notion of Chadbourne & Forster. A very small outlay of capital will fit up a tower in every respect superior to the one just sold. If there were lead enough, or demand enough, we could have a hundred shot manufactories at work here within two months. Who will be the man to move in the enterprise?

The paper finally encouraged local citizens by saying this shot-making experiment was just the tip of the iceberg and that more shot-making

Who needs the Shot Tower? When mineshafts like this one became abandoned, Dubuque entrepreneurs extended their height with surface platforms and produced shot by dropping molten lead down inexpensive "shot holes." *D.D. Owen, illustrator, Calvin Photographic Collection, University of Iowa.*

businesses would soon be seen in Dubuque. It also got in one more slap to Chadbourne & Forster:

> *The experimenting parties were very much elated by the result of their trial, and consider their discovery very valuable. They expect it will lead to important results, and look forward to the creation of shot manufacturing works here within a few weeks, which will compete with the world. If C&F want to buy up all our mineral holes, now is their time to do it.*

Many citizens must have read these articles because they began doing their own experiments to make shot. The *Daily Times* was glad to tell its readers the news and to let them know this process was already working in other states, and even in other countries:

> *MORE EXPERIMENTS—Experiments are daily being tried in the way of making shot by dropping it down mineral shafts, and with still more satisfactory results than at first. It is rumored that one of our citizens intends to embark in the enterprise immediately. It turns out that this method of making shot has been pursued with success for many years in Germany, and to some extent in Missouri, so the problem may be considered solved, that there is no necessity of building $15,000 towers when a hole in the ground, with an expenditure of $500, will do as well.*

According to an article that was published in the September 26, 1943 issue of the *Telegraph Herald*, Graves wasted no time in coming up with a successful way to make shot without a tower:

> *The purchasing agent* [for Chadbourne & Forster] *had hardly got aboard a South-bound boat, when Mr. Graves bought another complete set of shot producing machinery. While waiting for the equipment to arrive he did a bit of surveying in hills and bluffs surrounding the city, and discovered an abandoned 157-foot mine shaft near Asbury Road. He made a lateral incision into the side of the bluff, striking the shaft 100 feet below the surface, where he blasted out a room. He moved the machinery in and within three months was producing more shot than ever was produced by the old shot tower.*

Even though those St. Louis businessmen Chadbourne & Forster tried to shut down the Dubuque Shot Tower, it turns out that the crafty businessmen in Dubuque gave them the "shaft"!

Chapter 25
A Mysterious New Addition to the Shot Tower

By Zac Brothers

The Shot Tower in Dubuque stopped making shot in July 1862, when it was sold to the St. Louis shot-making company Chadbourne & Forster, which closed it down. Then, on August 26, 1871, the *Dubuque Daily Times* reported that the structure was sold one more time:

> *Sold for Delinquent Taxes—The old and conspicuous landmark, the Dubuque Shot Tower, which stands near the western terminus of the railroad bridge, was sold yesterday for delinquent taxes. The tower was purchased some years ago by a St. Louis shot company merely to prevent competition, and has been permitted to stand idle ever since. The company seems to have neglected to pay the taxes accruing upon the same, and hence the sale.*

About two years later, the *Dubuque Daily Herald* published an article about the Shot Tower, but not because of the amount of shot it used to make in the old days or because someone else had bought it. This time, the article of April 17, 1873, described how the tower was becoming a dangerous building where children playing in it might get injured. The paper made a suggestion as to how to fix the problem:

> *Venturesome boys are in the habit of playing in the Shot Tower and climb up the inside nearly to its top on old and decayed stairs of wood work liable*

> *to give way at any moment. The practice is a dangerous one and unless discontinued will result in an accident if not death. The windows of the shaft should be boarded up to prevent egress.*

From these newspaper accounts, it can be shown that the Shot Tower stood abandoned, falling apart and was a dangerous place for kids to play. One day, however, that all changed.

On the cold and wintry day of January 31, 1874, people in Dubuque woke up to find there was something strange and new on top of the Shot Tower. It appeared to be a statue of someone sitting on a horse, but it was hard to tell because the height of the tower made it difficult to see the figure's face. One curious and daring Dubuquer decided to enter the tower and climb to the top to get a better look at the statue. In 1970, a man named Harold Hennig wrote a short account of this incident. I found his article at the Center for Dubuque History at Loras College. He called it "The Mystery of Andrew Jackson's Statue." Hennig picks up the story just after the man came down from the top of the tower, and writes: "Upon

Andrew Jackson (1767–1845). Hero of the Battle of New Orleans in the War of 1812 and seventh U.S. president from 1829 to 1837. His wooden equestrian likeness sat atop the Shot Tower from 1874 to 1881.

his return, he faced the curious crowd and excitedly called out in a ringing voice; 'It's General Andrew Jackson!' A cheer for the revered, valorous old warrior went up from the crowd."

The very next day, the *Daily Times* reported, "No little surprise was evidenced in the city yesterday by those who happened to look toward the Shot Tower and saw mounted upon its summit a statue." On the same day, the *Daily Herald* published more information about the man who had something to do with the Shot Tower's new ornament and the man who created it:

> *John Deery is a public spirited citizen, and has adopted a singular way of displaying his enterprise. By means of subscriptions he secured enough money to rejuvenate Tom Kavanaugh's statute* [sic] *of Gen. Jackson, and his horse, which for years has stood sentinel at Mr. Kavanaugh's residence and mounted the wooden device on top of the shot tower, where it was placed yesterday with impressive ceremonies.*

Deery was a well-known attorney and Dubuque civic leader. Historian Jim Jacobsen adds that he was born in Ireland and came to Dubuque in 1854. He was the assistant editor of the newspaper called the *Northwest* and was postmaster during President James Buchanan's administration. (Buchanan was the country's fifteenth president and served from 1857 to 1861.) After that, he became a lawyer who specialized in real estate. Jacobsen said that Deery was "a lover of literature and art; he was the natural candidate to favor ornamenting the Shot Tower with an equestrian statue."

The day after the discovery of Jackson's statue on the top of the Shot Tower, the *Daily Times* printed a short story that said the Jackson statue had been in Dubuque citizen Thomas Kavanaugh's yard for over a year and that it was Mr. Kavanaugh who sculpted this equestrian figure out of wood: "The statue is the one that has been standing in Tom Kavanaugh's yard for a year or more past; is made of wood, and a very creditable specimen of the artistic skill of Mr. Kavanaugh. The stern old hero smiles down upon a mighty wide scope of country, commanding as he does an outlook of many miles."

Who was Thomas Kavanaugh? According to Jacobsen, he was quite a talented artist. In his research on the Shot Tower for the City of Dubuque, he included information from a *Telegraph Herald* article printed on October 10, 1915. Jacobsen said this about him:

Sculptor Kavanaugh was born in Ireland and gained his reputation for creating a statue of British Queen Victoria's Prince Albert. In Dubuque, he completed two equestrian statues of wood, reinforced with iron. The first, of George Washington, burned when the artist's house on Sumner Hill was destroyed. The second, of General Jackson, was intended for perhaps a temporary placement on the southeast corner of Jackson Park, opposite St. Patrick's Church. Kavanaugh was also the alderman for the First District.

It seems that John Deery hired a steeplejack (someone who is trained to work on tall structures and buildings) by the name of John Drehouse to lift Kavanaugh's sculpture up the wall and mount it on the flat surface on top of the Shot Tower. Jacobsen says that the carving was larger than the interior measurements of the building, so it had to be hoisted to the top from the outside.

Andrew Jackson, who was Kavanaugh's carving subject, was commander of American troops and became the hero of the Battle of New Orleans, where the British were defeated in the War of 1812. He was also the seventh president of the United States from 1829 to 1837. His nickname was "Old Hickory" because of his attitude of toughness. He also must have been greatly respected by Democrats, even though many years had passed since

This Cyril Ferring painting, commissioned in 1964 by the First National Bank to commemorate a century of business in Dubuque, is the only known image depicting the equestrian statue of General Jackson on the Shot Tower. *Dubuque County Historical Society.*

he was president, because the *Daily Times* on May 8, 1874, wrote: "The statue of Gen. Jackson, which surmounts the shot tower, is said to attract much attention from the boatmen; and those who are Democrats among them, especially, reverentially doff their hats to the old veteran who so calmly looks down upon them from his lofty eminence."

As time moved on, the weather in Iowa began to eat away at the Shot Tower and its oak statue. But perhaps because of the figure of Jackson sitting on his wooden horse, the tower became even more popular with local children. And just like today, some of those children back then took pocketknives and carved their names at the top. But first, they had to climb the creaking stairs to reach it. The November 11, 1879 edition of the *Daily Times* described the scene: "The old Shot Tower situated at the entrance of the bridge, is one of the old notable structures on the river. Almost every boy in the city knee-high-to-a-grasshopper has climbed to the top and cut his name on the wood with an old jack-knife or scratched it on one of the bricks."

The paper then reminded its readers that the Shot Tower had been sold in 1862 to the shot-making company of Chadbourne & Forster, which left it vacant and boarded up, and then described how dangerous the tower had become, even though kids were playing in it and around it with "Old Hickory" watching them:

> *Since that time* [July 1862] *it has stood vacant, and has been the roosting place for the inhabitants of the atmosphere and the look-out for those who dared scale its weak and rotten stairway. Yesterday a visit was paid to the old romantic weather-beaten pile of stone and brick. Some acrobatic variations, such as hanging by your heels, skinning the cat, etc., finally took us to the first story, after which it was fair sailing over broken doorsteps, etc. A single misstep, or a stair giving way would send us flying through the air, and in all probabilities would wake up dead and find Coroner Coakley sitting over the remains.*

The reporter then got back to his main goal—to climb to the top of the Shot Tower to see what was going on. He wrote, "But away with idle thoughts and to the top to shake hands with that old brick, General Jackson, who keeps an eye on the City of Dubuque and East Dubuque." The reporter said that although the view was great and the top of the Shot Tower would also be a good place to get inspired to write some poetry, he was there to see the names that had been carved into the wooden window frames and bricks of the tower and into the wooden statue of Andrew Jackson and his horse:

> *But our object was not to guy off the beauties of Nature, it was simply to take down a few of the names penciled and cut there years ago by boys, who are now young men. On the window sills can be read, "Charles Otley, Dave Robison, Frank Stout, Jim Welsh, Geo. Henion, Joe Morgan, Ed. Shankland, W. Vandever, George Blossom, George Gray, Dr. J. Peterson, James Wallis, Jo Linehan, Will Andrew, M. Tschirgi," and others.*

Those were the names of boys who climbed the tower in the years before the statue appeared. The reporter then continued with names of kids who came later. He called them the "younger crowd":

> *A younger crowd has had the pleasure of cutting Gen. Jackson and his horse. On the horse's neck can be read, "C. Russell, Wadsworth Davenport, C.D. Ham, G. Graves, L. Graves, A. O'Malley," and others. In the brick, "Ed. Peasley, A. Flick, H. Schunk, T. Rogers, W. Morrill, Stanley Robison and C. Bissell" have left their marks.*

The reporter ended his story with this interesting comment about the history of these names and what they represented: "It is a pleasure to visit this old autograph album and read over the names, for it calls back the good old school days that have gone, never to return."

After several more years passed, the carved wooden statue of Andrew Jackson and his horse began to fall apart because of Iowa's harsh weather conditions. Mr. C.J. Lesure, a businessman who bought the Shot Tower for $425 in May 1881, removed what was left of the crumbling sculpture. The tower was now going to be turned into a watchtower for his new lumber company, and a sturdy observation deck would soon be built on the top. But the story about Jackson's statue wasn't over yet. One major question that still needed to be answered was why did John Deery choose the roof of the Shot Tower as the place to pay John Drehouse to put Tom Kavanaugh's statue of Andrew Jackson? One day, we might get to the bottom of this puzzling issue by finding information in a diary or someone's personal journal, possibly passed down in a family from relatives who were there when it happened. Otherwise, this part of the story of General Andrew Jackson's statue and the Dubuque Shot Tower will remain a mystery.

Chapter 26

Alexander Simplot

By Doug Lang

Before photography was invented, people had to rely on the work of artists to gain knowledge of how things looked. A man named Alexander Simplot was one of those artists. He would one day become known around the world for his artistic ability, and today we are fortunate to have examples of his work so we can learn what our town and the Shot Tower looked like many years ago.

Simplot was born in Dubuque, Iowa, on January 5, 1837, one year after his parents, Henry Frederick Simplot and Susan Leclere, came to America from France. At this time, Dubuque was just getting its name and beginning to grow. As a child, Simplot developed an interest in drawing, something his parents did not appreciate or approve of. His family was so wealthy that he was the only boy to have a drum that was shipped up the river from St. Louis.

Because he grew up in Dubuque, the Mississippi River may have influenced his artwork. As an adult, he often drew pictures of the Civil War along the two rivers that border Iowa, the Mississippi and the Missouri. For example, when he was twenty-four, he created a sketch showing men who had volunteered to fight in the Civil War leaving Dubuque on a riverboat called the *Alhambra*. He sent this artwork to *Harper's Weekly*, a magazine company during this time. They must have liked what they saw, because later in 1861, *Harper's* hired him to draw pictures of battles and other events in the Civil

Alexander Simplot (1837–1914). One of Dubuque's original native sons whose sketches of Civil War scenes for *Harper's Weekly* earned him a national audience. After the war, Simplot created detailed drawings of Dubuque cityscapes. *Center for Dubuque History*.

War to put in their magazine. He was one of only thirty other artists who drew for *Harper's Weekly* during this conflict. Among those artists, he was the only one who drew pictures of the correspondents at work and at play. His work became known across the nation as readers bought the magazine to keep up with news of the war. However, a couple of years later, in 1863, he was forced to go back to his hometown of Dubuque because he developed dysentery, an inflammatory disorder of the intestine causing pain, fever and severe diarrhea.

In 1866, at the age of twenty-eight, he married an eighteen-year-old woman by the name of Virginia Knapp. When it comes to determining how many children Alexander Simplot had, the information is not all the same. It is somewhat of a mystery. For example, an article in the William Bowell Library at the National Mississippi River Museum said that he only had one son, Julien Dubuque Simplot, who lived into the 1950s. However, the internet website www.familytreemaker.com states that he had six children: Julien Dubuque Simplot, Ella Nora Simplot, Ada May Simplot, Mary Amanda Simplot, Alvin Reuben Simplot and Orville Gerald Simplot. The Center for Dubuque History has photos of Simplot's children, so I think that would be enough proof that Simplot did have a family in Dubuque.

At the end of his career of drawing for *Harper's Weekly*, the company had published fifty-one of Alexander Simplot's drawings in its magazine. If his artwork was good enough for *Harper's Weekly*, we can assume that his etchings of Dubuque with all its homes, businesses, bridges, roads and the

Shot Tower would be accurate accounts of what the riverfront area looked like at that time.

In one specific sketch, done in 1887, Simplot drew an image of Dubuque and the local surrounding area near the Mississippi River. When this etching is compared to photos of the riverfront in Dubuque today, Simplot's artwork shows that the southern area along the river, where the Grand Harbor Resort and Water Park is now, is all filled-in land, and the area east and north of the Shot Tower has also been filled in. This etching shows two bridges that go across the Mississippi River. One is the train bridge that was completed in December 1868, which is still there. The other is a wagon bridge, completed in November 1886. It was used by horse-drawn buggies and pedestrians. That bridge was taken down in 1943, shortly after the Julien Dubuque Bridge was completed. The Shot Tower is visible to the north of the two bridges in Simplot's sketch.

When photography became known and popular in the mid-1850s, many people in the United States turned to cameras for pictures instead of actual drawings or sketches. Photos were growing popular because they were more accurate and took less time to develop. However, Alexander Simplot chose

Simplot's postwar sketches documented local scenes. This image offers a view from East Dubuque that includes the railroad bridge, Mississippi River traffic and the Shot Tower in the distance. *Center for Dubuque History*.

Dedicated on October 31, 1897, the new Julien Dubuque Monument overlooks Catfish Creek, where the noted Frenchman mined lead with local Indians. Standing fourth from the right wearing a bowler hat is monument designer Alexander Simplot. *Center for Dubuque History.*

not to use photography but decided to continue to sketch his ideas and images. He and other artists kept creating drawings of the city of Dubuque. Some of these images show where the Shot Tower stood and what was nearby. These sketches can be compared to what is present today in Dubuque at those locations. They also provide an important link in the study of the Shot Tower's history in the 1800s.

Simplot had other talents as well. For example, he was the person who designed the limestone grave and monument that was made for Julien Dubuque that still stands today on the bluff that overlooks Catfish Creek. This grave site replaced the wooden grave the Indians made for Dubuque when he died in 1810. Simplot was also the secretary of the Old Settlers' Association in Dubuque.

On October 21, 1914, at age seventy-seven, Alexander Simplot died. He is remembered today because of his great Civil War artistic abilities, and during this time he was Dubuque's oldest native-born son. We see his work as important because it helps us to get a better idea of how the Shot Tower looked in the 1800s. He is buried in Linwood Cemetery in Dubuque, Iowa.

Chapter 27

The Fiery End of the Shot Tower

By Renee Anderson and Carol Cross

Editor's note: This chapter focuses on the damage sustained by the Dubuque Shot Tower during the devastating fires that destroyed the Standard Lumber Company in May 1911. It is not intended to be a detailed account of the blaze itself.

After the Shot Tower stopped producing shot in 1862, it was sold several times. In 1881, the owners of a new Dubuque business, the Standard Lumber Company, bought it and used it as a watchtower to protect its lumberyard from fires and theft. The historic structure had served its purpose for making shot, and now it had a new job. However, after being the lumberyard's lookout spot for thirty years, one day in Dubuque history changed the Shot Tower's entire reason for being.

Just before midnight on May 26, 1911, a fire was set at the Standard Lumber Company's lumberyard. It was discovered by an alert watchman, who sent an alarm to the central engine house at 12:15 a.m. on May 27. According to the *Dubuque Telegraph Herald*:

> *When the department arrived at the fire the blaze was already beyond any hope of control. The blaze seems to have originated near the old shot tower, lately used as a watch tower, but a fresh breeze from the south, had spread the fire along a considerable distance...So intense was the heat in which Number One company worked that their rubber coats were blistered into*

Standard Lumber Company's new watchtower, overlooking the extensive inventory of finished and stacked lumber in the yard on the left, with Lake Peosta in the background and the Mississippi River to the far right. *Dubuque County Historical Society.*

A close-up image of the Shot Tower featuring the enclosed observation deck built by the Standard Lumber Company, where watchmen protected from the elements could keep a lookout for fires and prevent theft. *Dubuque County Historical Society.*

> *uselessness and their heavy leather helmets charred so they could be picked to pieces with the fingers.*

Then another suspicious fire broke out about two blocks away:

> *While the fire was at its height, a second blaze was discovered, this also of incendiary origin, nearly two blocks away from the big fire and west of the Flick-Zieprecht box factory. Here a quantity of oil-soaked waste, such as is used in the bearing boxes of railroad cars for lubricating the journals, was found stuffed between the boards in a pile of lumber. Fortunately, the blaze was discovered before it had more than got started and was extinguished with a few buckets of water.*

The *Daily Times Journal* reported that "the police were at once notified of the discovery of the fire-bug's work and every detective and policeman in the entire department was notified to be on the lookout for the fiend."

According to historian Jim Jacobsen, over six city blocks of stored white pine lumber eventually went up in smoke. The *Telegraph Herald* added, "Most

Ground level at the Standard Lumber Yard, with the Shot Tower distant center. Finished lumber was stacked with spacers to air dry, which unintentionally contributed to the rapid spreading of fires once they were set. *Dubuque County Historical Society*.

of the area being devoted to piles of finished lumber, flooring, siding, etc., all of white pine and all tremendously valuable." The glow of this fire could be seen for over ten miles, and the total loss was around $400,000. Jacobsen writes that the estimated amount of material destroyed was equal to ten million board feet of lumber. (Editor's note: A representative from the Spahn & Rose Lumber Co. in Dubuque said this would be the equivalent to nearly four hundred semi truck loads of finished lumber. The financial loss would be the equivalent to nearly $9 million in 2010 dollars.) It was bad enough that so much lumber had been needlessly destroyed. What happened next would change how people looked at the Shot Tower for almost one hundred years.

There are not many building materials that are stronger and more resistant to fire than brick and stone. But even though the Dubuque Shot Tower was constructed with them, this lumberyard fire did the damage that made the Dubuque landmark virtually worthless and almost succeeded in destroying it once and for all. The *Telegraph Herald* explained in sad detail what happened to the Shot Tower as the flames from the lumberyard fire did their work. Notice that some think that just the intense heat of the blaze alone could have made the wood structure inside the Shot Tower and the observation deck explode into flames:

> *Perhaps the most spectacular feature of the fire was the burning of the old shot tower, a relic dating back almost beyond the memory of the oldest inhabitant…Saturday morning's fire leaves it still standing but nothing but a shell. Whether a spark or from the intense heat the flames burst first from the roof, high above the highest flames. Then the windows of the floor below caught and so down to the bottom. Finally, the interim woodwork caught* [fire] *and dense black smoke poured in a ribbon from the top to bottom of the tower. When this cleared away the burning timbers of the floors could be seen falling inside, shooting past the windows and sending up showers of sparks.*

Later on in the same article, the *Telegraph Herald* added to its description of the blazing Shot Tower:

> *The burning of the old shot-tower, which for years has been one of the best known land-marks of the Key City, was watched with a fascination on every hand. The structure is hollow, and as the windows fell out, the shaft of fire could be seen raging from top to bottom, belching forth from the top in a shower of flame which rendered to it the aspect of a mammoth sky-rocket.*

The blazing Shot Tower. Once its wooden observation deck and internal structure were destroyed by fire, the old landmark was rendered useless. *Center for Dubuque History*.

> *From out the sides, a solid wall of smoke waved in the breeze, reminding one of an immense flag afloat in the night and tinged with the conflagration which roared and hissed beneath it.*

The *Times Journal* described the horrible scene this way:

> *As the flames were roaring and crackling through the hundreds of piles of lumber, the shot tower, a landmark of the lumber district, was a mass of flames from top to bottom. This was probably the most spectacular feature of the big conflagration, as the tower is 150 feet high and flames and smoke were pouring from every window.*

As the fire got stronger, the Dubuque Fire Department tried very hard to put out the raging blaze: "The sullen crackling of the flames as they ate their course of destruction with fiendish fury could be heard for blocks, and the sight of the firemen, standing rigid with the hoses playing on the flames, their forms silhouetted against the brilliant wall of color, made up an incident long to be remembered."

The *Telegraph Herald* also recorded Dubuque fire chief Reinfried's observations about how the Shot Tower burned:

> *Speaking of incidents of the fire, the chief stated the really novel feature of the action of the fire was demonstrated in connection with the shot tower. The stone shaft caught fire from the top and burned down. The tower stands to the south of the origin of the blaze. The wind was from the south. Yet the intense heat from the burning pine when it reached the altitude of the top of the tower was such that it caught* [fire] *and the path of the flames was through the center of the structure and to the ground...Just how seriously the walls are injured and whether it will be necessary to level the structure which has stood for fifty years, has not been determined.*

A reporter for the *Times Journal* saw Dubuque's mayor, Daniel J. Haas, at the scene of the fire and asked him to comment on the tragedy. The mayor said:

> *This is simply awful. That such a fire should occur at this time is greatly to be regretted. I don't know anything about the case of the various fires about the city tonight, but I understand there are evidences of incendiary work. The police are busy on the case and will leave nothing undone that can be done to land the parties guilty of the crime.*

Who was responsible for this terrible tragedy in our city? The early evidence strongly pointed to a local man, as reported several days later in the May 30 edition of the *Times Journal*. Police picked up a young adult who lived and worked in town: "The police have arrested John Dixon, 22 years old, residing in the flats, and employed by the Adams Co., on suspicion of being the fire bug. The authorities absolutely refuse to discuss the arrest, but say they have the right man. The men arrested Sunday have been released. Dixon is known as 'Sleepy.'"

Even though the Standard Lumber Company had insurance that covered two-thirds of its damages, the business was declared a total loss. About a month after this fire, the charred Shot Tower was sold to M.H. McCarthy for one dollar on June 25. Due to the collapse of the floors and stairs inside it, and the destruction of the observation deck on top of it, the Shot Tower could not be used for a watchtower any longer. Now that its interior was burned and gone, it had no value to anyone, and it soon became forgotten and abandoned. Thanks to this terrible lumberyard fire of 1911, the once proud Dubuque Shot Tower was now just a burned-out home for birds and bats.

Chapter 28
SOS: Save Our Shot Tower!

By David Kenniker and Carol Cross

It sits there all by itself. To the east, the Mississippi keeps it company. It is the Dubuque Shot Tower. At one time, this structure was used to make shot. At the height of production in 1862, the Dubuque Shot Tower was a major producer of shot in the Midwest. After it was sold to a competitor, the Chadbourne & Forster shot-making company of St. Louis, Missouri, on July 14, 1862, the tower was never again to produce shot.

The Standard Lumber Company then used the structure for a watchtower from 1881 until a fire on May 27, 1911, ripped through it. The wooden stairs, all thirty-six windows and an observation deck on the top of the tower were destroyed. This fire caused the tower to become worthless for any further use.

There was almost no newspaper coverage of the Shot Tower through the 1920s, and there was very little during the early 1930s as well. But in the fall of 1932, the Dubuque Women's Club asked the City of Dubuque to take steps to preserve the deteriorating tower. The city was developing plans to turn the land around the Shot Tower into lots that could be sold to businesses, and before too many companies signed on, the Women's Club wanted to make sure the Shot Tower would be protected by creating a small park around it. They warned that "each year sees further deterioration of its masonry" and asked the city to tuckpoint the tower and put a temporary roof on it. As reported in the October 11, 1931 *Dubuque Telegraph Herald*, the Women's Club also recommended the installation of a beacon light on the tower:

> *The old Tower could be made into a beacon so that its beams could be seen for miles up and down the river and across the country into the three states, it would again have a real purpose in addition to its historic growth. In the event of an airport being established permanently here, the light would be of inestimable value.*

There had been ideas of putting some kind of light on top of the Shot Tower just before it was completed in 1856. Perhaps the Dubuque Women's Club knew that and wanted to see it finally come true.

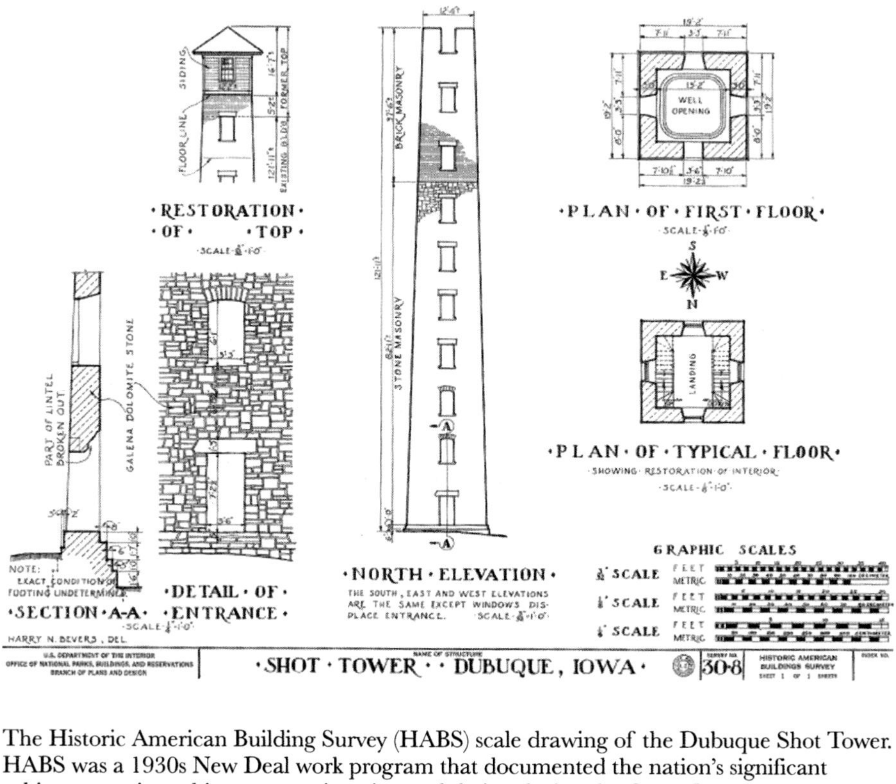

The Historic American Building Survey (HABS) scale drawing of the Dubuque Shot Tower. HABS was a 1930s New Deal work program that documented the nation's significant achievements in architecture, engineering and design during the Great Depression. *Dubuque County Historical Society.*

Then on August 6, 1933, the *Telegraph Herald* ran a story that celebrated the Shot Tower's seventy-fifth birthday, even though it was actually seventy-seven years old by then. The same article said the Shot Tower had also been heavily damaged by another lumberyard fire in 1894 as well as the fire of 1911. In the 1930s, the country was suffering through the Great Depression. To help with the high unemployment rate in the country, a make-work project called the Historic American Buildings Survey program was created in 1934 to give out-of-work engineers and architects jobs and also contribute something to the country. The Shot Tower was selected for documentation by this group. It was measured and evaluated, and detailed scale drawings were created to record its existence as a historic American building. A brief history of the tower was also provided in the final HABS report. However, no actual work was done to help repair or strengthen the Shot Tower.

Even though the Dubuque Women's Club came up with some very good suggestions for improving the Shot Tower, nothing was done to it, and the years just passed by. Then, about ten years later, on September 26, 1943, the *Telegraph Herald* called the Dubuque Shot Tower a "famed local landmark" and noted that it had "played a very important role in the prosecution of the Civil War." The paper then added, "It has stood as a silent tribute to Iowa Union veterans in three wars, the Spanish American conflict in 1898–99, and the two world wars." By this time the Shot Tower was getting close to its 100th birthday, and this made it easier for people to begin to consider accepting it as a landmark.

After World War II, the Shot Tower became an Iowa icon because of an article that appeared in the *Palimpsest*, the state historical society publication. Dubuquer "Steamboat" Bill Peterson was a supporter of the Shot Tower's protection and improvement ideas. The article he wrote for the *Palimpsest* stressed the role of the tower as being a survivor of Dubuque's history.

On January 2, 1949, the *Des Moines Register* called the Shot Tower a "long-neglected historical monument," and on December 25, 1949, the *Telegraph Herald* reported that local Girl Scouts had an idea to plant shrubbery around the tower's base if the city would clean up the area, but nothing happened to improve the site. A few years later in 1951, crazy claims about the Shot Tower got spread around when *Railroader Magazine* published a photograph of the tower on its cover with the caption that said it had "furnished much of the shot used by the Union Army during the Civil War." However, this was an inaccurate statement. Also in 1951, industrial buildings in Dubuque were being constructed just north of the Shot Tower, and an engineer who was evaluating the land to see if it could support new building foundations said that his opinion was that the famous structure could easily last another one hundred years. He said that its mortar was "as hard as flint" and "the stone is made of the hardest found in this vicinity." A few rocks had fallen from the first-floor window frames, "which is now level with the surrounding terrain because of filling." A few stones had also fallen from the upper stone arches, but the Shot Tower seemed to be a very solid structure that would stick around for a long, long time.

The Dubuque Chamber of Commerce declared its support for doing something to improve the Shot Tower in 1952 "because of its great historical interest" and its "trademark" role for the city. The beautification committee of the chamber studied the issue and said that something had to be done to improve the tower's condition "before it's too late." The city engineer studied the Shot Tower and developed a $7,000 project budget. One major concern was all the loose bricks on the top of the tower's walls. If they fell to the ground, these bricks could seriously hurt any children playing around the tower and damage passing trucks. People were also fearful that an out-of-control truck could strike the Shot Tower and bring it down "like a house of cards." The chamber's beautification committee handed out its report in 1953 and made a suggestion that the tower's lower openings should be sealed off. It also said that a new roof should be installed. The *Telegraph Herald* reported that raising a sum of $5,000 to $15,000 would be accomplished by creating a fundraising drive that would include the community's schoolchildren. The first real Shot Tower restoration work was discussed in 1956, and estimates to improve the building, including tuckpointing, replacing concrete sills in all the windows and adding a new concrete roof, were about $7,500.

By 1959, the Dubuque County Historical Society, led by a woman named Mrs. Abbie McDonald-Dancer, was taking the lead in saving the Shot Tower. According to the October 25, 1959 edition of the *Telegraph Herald*, it seemed that the tower was now just about ready to fall over, even though people had said in 1951 that it would be able to stand another one hundred years. A local contractor warned that the stonework was merely a "veneer" and that there were "plenty of holes" in the tower. Mrs. McDonald-Dancer said that the wear and tear was occurring so fast that the historical society was even thinking about getting the Shot Tower insured against collapse. McDonald-Dancer and her group then joined with the local vacation committee, other Dubuque organizations and private citizens and worked to support the idea of getting the Shot Tower restored.

Back in 1950, the City Council had budgeted $5,000 for the Shot Tower's repair work, but the Dubuque County Historical Society stated that a total of $12,000 would be needed. By this time, the tower was called "the oldest landmark in this part of the country." Then the *Telegraph Herald* reported on October 25, 1959, that the tower was "the only real symbol of Dubuque." In November 1959, a special Save Our Shot Tower Committee, chaired by Henry Miller, was created. This group of local citizens decided to help complete the earlier plans of the city to restore life to the tower. The committee members included Elmer Brauhn, David Hammer, Frank Hardy, Charles Scherrer, Nick Schrup, Joseph Wenzel, William Hill, Donald Huntoon and M.S. Engelbrecht. Mr. Miller, who was also a member of the Dubuque County Historical Society, said, "There is a real urgency about the situation. If something isn't done soon, the Tower will fall down. That's all there is to it." Miller went on to say, "The public has to become interested in the project. Interested enough to lend support." Mr. Miller added that the Dubuque Chamber of Commerce already had said it would give "substantial and continuing" support to the tower if the citizens of Dubuque showed they were interested, too. "Of course, the project is still in the planning stage, but with the public behind it, it can become a reality," he said.

During November 1959, the Save Our Shot Tower Committee began its fundraising campaign, and money was raised from private donations and local businesses. With the help of American Bank and Trust (today it's called the American Trust and Savings Bank), the tower got its own trust fund account. On November 22, 1959, according to the *Telegraph Herald*, coffee cans went out to local businesses, and the public started to contribute with their spare change and dollar bills. The SOS activity got a real boost when the famous broadcaster Paul Harvey talked about the Dubuque Shot Tower on his radio show. He was heard on more than 470 radio stations across the country at that time, so the Shot Tower instantly became very famous. There was even a Sunday event held at the St. Elias the Prophet Greek Orthodox Church in Dubuque, and $1,000 was raised in a special "Save the Shot Tower" dinner.

Historian Jim Jacobsen reported that the SOS committee set a goal of $7,000 to help fix up the Shot Tower and finally raised $6,848 after the goal was reduced to $6,700 by late March 1960. The restoration plan by this time was more expensive and included more repairs and improvements to protect the Shot Tower from the weather. The Shot Tower would be tuckpointed, the area around the structure was to be lighted and cleaned up and there

SOS—This "Save Our Shot Tower" broadside was circulated by American Trust and Savings Bank of Dubuque to promote community fundraising efforts for the Shot Tower's rehabilitation project in 1960. *Dubuque County Historical Society.*

were even plans that lead shot-making demonstrations would be made at the tower. Tours would also be offered. Then the real plans to fix up the Shot Tower started to come to light. The process was started to get bids from interested companies so the restoration work could get underway. According to the *Telegraph Herald* of May 17, 1960, the company that was awarded the contract was the George Jackson Steeple Jacking and Tuckpointing Company of Dubuque. Its fee was $11,000 to waterproof the structure and put a cement slab on the top of the tower.

The Jackson Company said that the Shot Tower was in reasonably good condition, even if some people feared the tower could fall over at any time. Tuckpointing on the interior and exterior of the Shot Tower was scheduled. They wanted to cut out all the old mortar and replace it since the inside was totally ignored. All cracks and spaces would have concrete poured in to fill them up. New stone louvers would be put on the first three floors. The new window framing was to be made of wood, but it was later changed to concrete to help protect the inside walls and the small cracks. Four sets of tie rods and cast-iron star bolt anchors were added to the outer walls of the tower. These sets were added at the fifth, seventh and ninth floors, tying them together going east and west to reinforce the Shot Tower against collapse. A concrete cap was installed on the top of the tower where the observation deck used to be to keep the weather out. A new concrete base was also poured around the tower's base. All of this work was completed in 1960. The interior walls were tuckpointed, and concrete lintels were put in place in the windows. The roof had lightning rods installed on it for safety in case the tower was struck in stormy weather. Even though none of the tourism improvements were done, the Dubuque County Historical Society then reopened the Shot Tower to the public in November 1961.

In 1967, the famous Shot Tower site was finally protected from any more flooding when a city floodwall was built by it. Then the Dubuque Garden Club landscaped the base of the tower. In addition, an iron fence was installed on an oval stone base that surrounded the Shot Tower and acted like a protective barrier. This was funded by private service groups. In 1969, a bronze plaque was placed on the east side wall of the structure by the Dubuque Chapter of the Daughters of the American Revolution.

By this time in its long history, the citizens of Dubuque at least had begun to make changes in the Shot Tower to improve its chances of staying in one piece, being able to survive the harsh weather of Iowa and maybe even stay around for another one hundred years. It wasn't a perfect solution, but at least it was a beginning.

Chapter 29

Other Shot Towers That Still Stand

By Doug Lang

The George W. Rogers & Co. Shot Tower located in Dubuque, Iowa, is unique in a couple of ways. First, its architectural style and shape, like old Egyptian obelisks, is very identifiable. Many other shot towers were built with rounded surfaces, looking more like tall smokestacks. And not many shot towers had the number of windows the Dubuque Shot Tower has. Second, it is the only remaining shot tower in the United States that is found west of the Mississippi River. However, even though it is a unique structure, it is not alone. There are six other towers in various parts of the country that you can still see today.

The Jackson Ferry Shot Tower

The Jackson Ferry Shot Tower, built in 1807 by an English immigrant by the name of Thomas Jackson, is located in Wytheville, Virginia. This structure was built out of limestone, with walls that are two-and-a-half-feet thick, which helped keep the tower's inside temperature constant and improved the quality of the shot made there. This tower is 75 feet high and sits on top of a 75-foot shaft, which makes the total drop distance 150 feet. It is the only known shot tower in the United States with both the shaft and part of the tower extending above ground level. It was added to the National Register of Historic Places in 1969.

THE SPARKS SHOT TOWER

The Sparks Shot Tower, located in Philadelphia, Pennsylvania, was built by two plumbers, Thomas Sparks and James Clement. This shot tower was built in America in 1808, just after the United States passed the Embargo of 1807 that stopped trade with England, which included shot that was made there. The Sparks family kept it running for four generations. It produced shot from 1812 to 1903. Shot used in the War of 1812 and the Civil War was made in this particular tower. The structure is 150 feet in height and is made of brick. Its round base has a 30-foot diameter, and it tapers to 15 feet at the very top.

THE PHOENIX SHOT TOWER

The Phoenix Shot Tower is the last one of four towers built in Baltimore, Maryland, that is still standing. It has also been called the Merchants' Shot Tower and the old Baltimore Shot Tower. This structure was built in 1828, and over one million bricks were used to make it. Also, it was constructed with no external scaffolding to help the workers. At one time this was the nation's tallest shot tower, as it stood over 234 feet high. In fact, for years, it was the tallest structure in the United States, until the Washington Monument was built after the Civil War. At one time it was one of the country's largest ammunition suppliers, producing 500,000 twenty-five-pound sacks of shot per year. In 1892, shot-making operations were ended. In 1972, the Phoenix Shot Tower became a National Historic Landmark, and in 1976 it was opened up as a museum.

THE TOWER HILL SHOT TOWER

The Tower Hill Shot Tower is located in Spring Green, Wisconsin, and was a tower our class visited to learn how shot was made there. This tower is a little different from other towers because it is made of wood and stone. A 60-foot wooden shaft was built on top of a 120-foot shaft that was cut into the stone bluff on the Wisconsin River, making the drop a total of 180 feet. There is a 90-foot horizontal tunnel, called an adit, that was dug into the bluff that allowed workers to walk from the riverbank into the stone formation and retrieve the cooled shot. This tower produced between six hundred and eight hundred pounds of shot per day. The smelting house was located at the top of the shaft, where shot of different sizes was made. This tower's shaft into the bluff was hand dug by Daniel Whitney and T.B. Shaunce. These men began construction of this tower in 1831, but the Black Hawk War of 1832 forced them to stop digging. The tower was then finished in 1833, a year after the war ended. It closed for good in 1860, just before the Civil War began.

The Remington Shot Tower

The Remington Shot Tower was built in 1909 in Bridgeport, Connecticut. This particular tower stands ten stories tall and is made of brick. It has a 133-foot drop with water tanks at the base. During World War I, shot production at this tower was greatly increased to keep up with the demand for ammunition.

The Winchester Shot Tower

The Winchester Shot Tower is still in operation today in New Haven, Connecticut. Some sources claim it was constructed in 1915. However, a video documentary called *Hands on History* claims it was built in 1903. This tower is 182 feet tall, and when shot is made, it is dropped eleven stories and uses the process very similar to the one patented by William Watts in Bristol, England, in the late 1700s.

Chapter 30
Joseph Pickett Sr.: A Man on a Mission

By Samantha Feyen

The following information comes from "Look Out Below," an article published in the *Dubuque Telegraph Herald* on October 17, 1976, written by Calvin Gatch Jr. He explained how a local citizen experimented with making shot from the Shot Tower to celebrate America's 200th birthday in the U.S. Bicentennial. After it closed in 1862, the Shot Tower was no longer used for its original purpose of shot production. Gatch told readers that the Shot Tower is located in downtown Dubuque, on the corner of Seventh and Commercial Streets. It is one of the last standing shot towers in the United States. It stood abandoned for years with burnt and charred interior walls and boarded-up windows, until one man with an interesting idea entered its future.

In 1971, local brew master, Joseph Pickett Sr., from Dubuque, reopened the Star Brewery, located just south of the Shot Tower. Before Pickett purchased the brewery, his friend George Forby Jr. gave him the idea of making shot by reproducing a manufacturing process that hadn't been tried in 114 years. Forby was a former Dubuque County Fair Association member who also happened to be a muzzleloading gun fan. Gatch reported, "The old muzzle-loaders, of course, were the weapons that pioneers used to fire shot." Pickett was sixty-seven years old at the time. He always knew that one day he would try to drop shot in the Shot Tower, but he was not sure when that would be, so he tried it first at his brewery.

Gatch reported that making shot has what many people think are easy steps. Molten lead is dropped through a sieve one hundred feet in the air above a water tank. However, it took Pickett eighteen months before he even got a little close to finding the right combination of ingredients and the correct steps to making shot. It seems there was a lot more to the process than just melting lead and dropping it down a shaft into a tub of water.

According to Des Moines historian Jim Jacobsen, Pickett tried to drop shot at his own brewery production site with a beer barrel full of water as a cooling tank for the lead to land in after being dropped from the fourth floor. However, wind blowing in through the brewery's windows plastered the brick walls on the inside with molten lead. After that, Pickett then tried to drop hot liquid lead from the top of an elevator shaft in the brewery. When even *that* failed, he became determined to use the Shot Tower for his experiment. He simply was not going to give up.

Gatch also told readers that in 1911, the lumberyard that used to be located by the Shot Tower was set on fire by John "Sleepy" Dickson. The wooden observation deck on the top of the Shot Tower caught on fire at that time, and it eventually burned the wooden steps inside from top to bottom. Some sixty-five years later, this made Pickett's job of getting to the top of the Shot Tower a huge challenge.

Pickett had to figure out a way to get into the Shot Tower and climb to the top. According to Gatch, Joe tried to get a Dubuque Fire Department truck to lift him up there. He also explored the option of using local cranes, but that would have cost him $150 an hour. He even looked into hiring a National Guard helicopter to fly him to the top of the tower, but because the guard pilots were unwilling to risk getting tangled in the electrical wires that were nearby, this plan was declined. Unfortunately, none of the ideas that Pickett came up with worked. However, he finally resorted to the use of scaffolding inside the tower. The *Telegraph Herald* reported, "The scaffolding was installed inside the Shot Tower when Pickett persuaded tuckpointing company owner Fred Jackson to assist him in the project." Mr. Jackson provided the necessary equipment for free. Pickett was also helped by Richard Vermeer, a surgeon from Davenport, Iowa, who had met Pickett while sightseeing on the river and became interested in his idea to make shot at the Shot Tower.

On the day of the historic experiment, the *Telegraph Herald* reporter and photographer, Pickett and Vermeer all struggled to get comfortable on top of the Shot Tower. Pickett cracked, "You can tango up here but there is not enough room to do the waltz!" Also on the four- by six-foot platform

After much planning and experimentation, local brewer Joseph Pickett Sr. and Davenport surgeon Richard Vermeer succeeded in producing lead shot at the Shot Tower in 1976. Said Pickett, "Making shot, like brewing, is an art." *Dubuque Telegraph Herald.*

were the materials needed to make shot: a couple of plumber's acetylene furnaces owned by Buechel Plumbing Co. and two of Pickett's beer boxes, overflowing with heavy leather gloves, frying pans, pieces of assorted lead and iron ladles, among other items.

When Pickett first melted and dropped lead at the brewery with his friend John Burns, many of the pellets took on the shapes of little gourds and pancakes. But Pickett said, "Making shot, like brewing, is an art. And like brewing, there are a lot of secrets we couldn't find out about from books." According to the *Telegraph Herald* article, Joseph Pickett discussed several factors that he learned were required for good-quality shot to be produced.

First, he tried different kinds of lead, including some found in the Dubuque area, but they didn't work as well as the lead he got from Leeds, Alabama. The temperature of the lead, the iron ladle and the sieve with holes where the shot is poured and dropped all must be "just right." If the temperature is too hot, the lead pellets will "explode" when they hit the water in the tank at the bottom of the Shot Tower and wind up looking like pancakes. Pickett took the temperature of the lead he melted by sticking a small piece of firewood in it. He found that when the wood got charred, the lead was the right temperature.

Second, there was a need for essential equipment. Pickett first tried to pour lead through a coffee strainer and through other kinds of mesh with holes. He even tried using a steel frying pan that he drilled full of holes. That was the idea that worked the best.

Third, he needed the right height. Pickett experimented making lead in his brewery's elevator shaft, which made a fall of about seventy feet. This method was easy because he could then pull the shot up to the top using a pulley. But it was not a long enough distance for the liquid lead to drop and form into round balls before it hit the water in the catch basin. It was at this point that he realized he needed more height.

Over time, Pickett's shot-making efforts transformed "meteorites" into spheres. The *Telegraph Herald* article also reported that he planned to make enough lead to fill about one thousand little airline liquor bottles with lead dropped from the historic Shot Tower for the Dubuque County Historical Society to use as fundraising items for the country's bicentennial. Unfortunately, after thirty years, many of the bottles that he produced have disappeared. With them, the information printed on small tags attached to the bottles has also been lost to history—or so we thought.

A man named Wally Krawczuk was a substitute teacher who worked at Central Alternative High School whenever a regular Central teacher got sick. One day he was at Central and learned about the Shot Tower class. He asked if he could make a presentation, and we are glad he did! No one had said anything to him about the little bottles that Joe Pickett filled with shot, because we figured they were gone forever and would never be seen again. But believe it or not, when Wally came to our class, he took out of a paper bag two original bottles of Pickett's shot! Now we had in our hands the actual shot that Joseph Pickett had made from the Shot Tower in 1976, and one of the bottles still had its original label tied to it. This was a tremendous find for us. Now we could tell people the whole story. The cardboard labels that Pickett prepared and attached to his bottles read as follows:

> *Lead mining was the major industry in Dubuque's early days. Bars of lead, called pigs were shipped by boat to St. Louis for conversion into shot. In 1856, a nine-story, 150-foot tower was erected by a group of Dubuque businessmen. From the top of the tower the molten lead was poured through screens of various gauges. The pellets thus formed, fell into cold water at the base of the tower. When cooled, these pellets were polished, graded and bagged for market. The tower operated successfully for about three years before the St. Louis firm lowered retail prices and made operation here unprofitable. The tower has stood idle since that time, one hundred and seventeen years. Mr. Joseph Pickett, whose brewery (the only such operation in Iowa) stands adjacent to the tower, conceived the idea of producing shot*

Souvenir bottles containing the first lead shot produced at the Shot Tower in 114 years since original operations ceased in 1862 were made for a Dubuque County Historical Society fundraiser and to celebrate America's bicentennial. *Dubuque Telegraph Herald.*

> *in the tower as a Bicentennial project. Scaffolding was built inside the tower, equipment and lead were acquired, but the challenge of rediscovering the formula for making the shot entailed many months of much patient work. After several failures, Mr. Pickett developed his now secret formula and produced the shot contained in this bottle.*

In 1976, the Dubuque Shot Tower was listed on the National Register of Historic Places by local historian Helen Mercer, so it was perfect timing for Mr. Pickett to make shot for the bicentennial. However, he told *Telegraph Herald* reporter Gatch that he didn't expect his shot-making time to last nearly as long as it had when the Shot Tower was at full production in the late 1850s and 1860s. The article ended with this quote from Joseph Pickett Sr., a man on a mission, about his attempt to make shot in the Shot Tower: "This will never happen again in another one hundred and twenty years because there aren't that many goofy people in the world."

Chapter 31
All Talk, No Work

By Johnna Foutty

In the late 1980s, there was some talk about what to do to improve the Shot Tower. However, that's all it was—just talk. The *Dubuque Telegraph Herald* reported on March 23, 1987, that there were several ideas that people were considering. The first idea was to move the 133-year-old structure to the Ice Harbor. Here is what the article said about the idea that was developed in 1986 by the University of Wisconsin:

> *A proposal to move the Shot Tower to the Ice Harbor came from the 1986 market and development study the University of Wisconsin conducted for the greater Dubuque Development Corps' Ice Harbor Committee. The study proposed the Shot Tower could serve as "the most viable option" for an architectural focal point at the Ice Harbor, and would "increase the public's exposure to the history it represents."*

However, nothing came of this idea. The *Telegraph Herald* reported that Jerry Enzler, who was then the Dubuque County Historical Society's museum director, said the idea of moving the Shot Tower would be pretty radical and probably not possible. The paper told its readers: "Enzler… said proposals to move the Shot Tower are both controversial and unlikely to happen. Controversial because such a project would remove it from its historical setting. Unlikely because of the cost and the fact that some Ice Harbor committee members consider it far-fetched."

The second idea was to actually cut $27,500 from the city budget and not bother to tuckpoint the city-owned Shot Tower. Public works director Frank Murray was quoted as saying that the Shot Tower's mortar showed little evidence of deterioration at the time but that he and his staff would reevaluate the tower's tuckpointing make-over and consider doing it in the future with another budget for improvement. The article also reported that in the 1980s, the City of Dubuque maintained the area that is now fenced in around the Shot Tower by mowing the grass and controlling the weeds. However, public works director Murray said the area around the Shot Tower was "not conducive to landscaping or other improvements because of adjacent truck traffic and the likelihood of vandalism in the isolated area."

The third idea, which was one that the Dubuque Jaycees were considering, was to place fencing to the north side of the Shot Tower and to install an informational kiosk. The article stated, "The club will decide within one to two weeks whether to proceed with the project." Of course, we all know that nothing was done. So again, all talk, no work.

Jerry Enzler saw the Shot Tower eventually tying into the Ice Harbor development, while the tourism activities spread across the Fourth Street peninsula toward the old Star Brewery and the tower. The above ideas that the Jaycees were proposing would be the only ones that were offered to improve the site at the Shot Tower. Basically, what this article was really saying was that many people thought about improving the structure, but it was only their thoughts that were making the Shot Tower what it was then and is today (2005)—an old and dirty structure. The history behind the Shot Tower is magnificent because of all it has been through over the years, and now it is one of the oldest, if not the oldest, structure still standing in Dubuque.

Although its history is outstanding, the appearance of the Dubuque Shot Tower was neglected. Sure, there was talk about restoring the great, tall building, but nothing was done. So up to this point in time, it was just an old, deteriorating brick and stone tower that was filled with debris and bird feces, mainly because there was all talk and no work.

Chapter 32
Hard as a Rock

By Johnna Foutty

The Dubuque Shot Tower stands tall and seems to be invincible. We know that the Shot Tower has been in flames and has even been beaten down by Mother Nature, but there is one thing that makes many people believe that it is really hard as a rock. On November 18, 1991, an article in the *Dubuque Telegraph Herald* explained how our beloved tower had been struck by a train car just three days earlier. This caused between $5,000 and $10,000 in damage to the structure. The remarkable thing is that the 135-year-old landmark appeared to show no signs of falling. The article explained how the accident occurred and when: "Empty grain cars were being switched on the tracks at the Peavey Grain Co. at East 7th and Commercial Streets about 10 p.m. Friday when the accident happened, police said. The train car is owned by the Chicago Central and Pacific Railroad."

Now we had some information, but where was the rest of the story? Readers had to wait over five months to learn more about this terrible accident. On April 6, 1992, the *Telegraph Herald* ran another article about the accident that damaged the famous Shot Tower by the Mississippi River. This account told how the city cleaned up the railroad car damage: "Mike Koch, Dubuque's public works director/city engineer, said the train car wreckage was removed from the site and a hole knocked out of the structure by the car was temporarily repaired."

Basically, the train car was moved out of the way and the hole was covered up with a piece of plywood. Here are a few words from the article as Mike

Koch explained the repairs and what precautions would have to be taken to protect the Shot Tower:

> *The city has drawn up specifications for the repairs needed to restore the tower and some protective fencing to its condition before the accident. The city expects to let bids for the project later this month, Koch said. "It* [the repair work] *shouldn't be too expensive, but the contractor will have to make the repair look like part of the historic structure. It may cost additional money to do that."*

The plan was to restore the Shot Tower back to the way it was before the train wreckage. Thankfully, the repair work was done so carefully and so well that today, most people can't even tell that such an accident ever damaged the structure. In a letter to the editor published in the *Telegraph Herald* on April 18, 1992, local citizen Mike Schmalz spoke his mind about the Shot Tower's repairs and who he thought should have to pay for it: "If a Chicago Central and Pacific train hit the old Shot Tower, why is the city responsible for fixing it? I wouldn't like to see a landmark like the tower be lost for something so stupid."

The City of Dubuque had the money to repair the Shot Tower, but in the end, the Chicago Central & Pacific Railroad did pay the cost for putting the Shot Tower back together again. The last time the Shot Tower underwent any major repairs was in 1959–60, so this was something that hadn't really been done in the past. Even though it got smacked by a railroad car, it is encouraging that the Shot Tower still stands today. It is an amazing landmark and will be for years to come. It's hard as a rock!

Chapter 33

Tourist Attraction

By Johnna Foutty

Our beloved Shot Tower. It holds many mysteries that we may never find the answers to. The most that people can do presently is go to the tower and observe it. One editorial in the *Dubuque Telegraph Herald*, published on April 10, 1992, argued that it was now time to "consider the Shot Tower" as an example of "untapped potential" and make it possible to be included with the city's other "architectural and historic treasures." The editorial headline said loudly, "Shot Tower: Time to tap its tourism potential!" The editorial put its case before the readers:

> *The 150-foot tower of limestone and brick is, at 136 years, the oldest structure on Dubuque's riverfront. It has presided over the city's transition from a newly incorporated municipality (Dubuque was only in its second decade when the Shot Tower was built) to an established and important Iowa community preparing for the 21st century.*

The paper argued that visitors to Dubuque easily should be able to see and enjoy this historic landmark. But the paper also said, unfortunately, "Few historic structures were built with history or visitor convenience in mind. The Shot Tower was no exception." Back when it was built, no one was thinking 150 years into the future that a business that made shot should also one day become a place for tourists to visit. Surrounded by industrial warehouse buildings, railroad tracks and weeds, the Shot Tower's surroundings are not pleasant. In fact, they are uninviting and ugly. The *Telegraph Herald* believed this all needed to change.

The article also said that with the repairs that were being planned to the tower, the City of Dubuque should include some ideas to improve the area around the Shot Tower to make it more appealing to visitors and tourists. It may require a public-private partnership, perhaps similar to the city hall clock tower project or even the drive of 1959–60 that kept the Shot Tower from crumbling in disrepair. The editors also reminded their readers of the crazy idea to move the Shot Tower to the Ice Harbor but said the decision to not go ahead with it was just as well because "it would have been a great structural and financial risk and would have diminished its historical significance."

In another section of this 1992 paper, readers were asked to phone in their opinions on what the City of Dubuque should do with the Shot Tower. The question of the day was: Should the City of Dubuque make repairs and improvements to the Shot Tower to increase tourism opportunities? The responses to this question were amazing. Of the people who called in, 76 percent said yes. Only 24 percent said, "No, they should not." Altogether, there were 148 calls to the *Telegraph Herald* about the improvements to the Shot Tower.

Why would people not want the Shot Tower to be fixed up? Possibly because they don't know all the history that is held inside that amazing structure. It made shot in our town in the mid-1800s. It was a watchtower for a lumberyard. It has survived a fire that burned down the steps inside. It has been struck by a train car, yet it is still standing to this day. What is even more amazing is that it is the only shot tower still existing west of the Mississippi River. If only everyone could come together and help make the Shot Tower more appealing, this structure would become known for its great history instead of for what it appears to be today—just an old pile of bricks and stone.

Chapter 34

The Shot Tower Restoration: Funding Sources and Their History

By Drew Tatum

Occasionally throughout its 150-year history, the Shot Tower has caught the attention of the citizens of Dubuque. In 2000, for example, a group called the Old Home Enthusiasts gave the City of Dubuque money to do a structural assessment of the tower. The city then hired the architectural consulting firm IIW (Iowa, Illinois, Wisconsin), which put Ron Balmer, the head engineer, and his crew in charge of checking out the tower. Ron and his associates came up with three different estimates for the restoration of the Shot Tower that varied from $225,000 to $1,840,000. They gave the city these restoration ideas and explained how much each of them would cost:

Restoration Idea #1: $225,000

A simple stabilization of the existing Shot Tower as it stands today with interpretive signs on the River Walk Project. This means that this proposal called for no public access to the tower but would remove all the bird droppings and refuse from the inside of the structure, patch up the cracks in the building and change out the cement lentils on the windows. There would also be permanent signs with historical information about the tower placed on the River Walk so tourists and local citizens could learn about it.

RESTORATION IDEA #2: $1,630,000

A "nominal circa 1856 restoration" would include minimal excavation and the addition of an "at grade" interpretive building, with a connection to the River Walk Project with public access. Removing bird droppings, conducting an archaeological investigation around and inside of the Shot Tower, installing new limestone sills and screening off all openings to prevent birds and other animals from getting inside were also part of the recommendation. This was different from the first proposal because of the excavation option and the new limestone sills.

RESTORATION IDEA #3: $1,840,000

This would pay for a more detailed restoration. The same work would be done as in the second suggestion, but there would be a lot more attention paid to the Shot Tower itself. This would include replacing damaged brick and limestone, removing loose materials from the interior walls of the tower and installing a fireproof stairway inside the tower that would extend to a rooftop observation platform. IIW would also favor constructing a 2,050-square-foot wood-framed interpretive building right next to the tower that would look very similar to the original outbuilding, as seen in a 1858 lithograph. This building would be fully air conditioned and would provide sewer service with ejection pumps and forced piping to keep water from building up by the tower. This would be the Cadillac of restorations. It would also involve having tie rods installed on the Shot Tower, which would help stabilize the walls, and complete plumbing plus complete excavation of the surrounding land.

The City of Dubuque looked over the three restoration ideas. Instead of choosing one, a mixture of the three ideas was agreed upon. This new option included the construction of a metal stairway and an observation deck. More importantly, the tower would be made available to the public. A slimmer budget was required. It was to be no more than $600,000.

The City of Dubuque then had to come up with a way to find enough money to fund the restoration of the Shot Tower. They first needed to have local money together before they could ask for a grant from the federal or state government. The funding by the city came in through private donations and from the America's River Project, a ninety-acre project at the Port of Dubuque that will create over one thousand new jobs. The city committed $200,000 as part of this program. The America's River Project was funded by the state. Altogether, they have given over $180 million toward working on Dubuque's Riverfront Project. After the city had the funds, they filled out an application for the federal and state grants because they would have matching funds. The city told them they had the funds needed to match the funds. The state grant came in first.

The State of Iowa awarded Dubuque $100,000 from the Historical Site Preservation Grant for maintaining general upkeep of the Shot Tower. Historical Site Preservation Grants can't exceed $100,000 and can be as small as $40,000. After receiving this grant, the city cannot apply again for three years. There also can't be two historical projects in one county.

In February 2003, the National Park Service approved the city's application to change the listing of the Shot Tower in the National Register of Historical Places to having national importance. The following year, in October 2004, Dubuque was given $295,000 from a federal grant called Save America's Treasures for the preservation of the Shot Tower. In order for the city to receive this grant, its project must already be funded by the Department of the Interior. The grant was created for cities and towns to preserve significant local structures that also have national importance, like the Shot Tower.

To receive the Save America's Treasures Grant, Laura Carstens, planning services manager, had to fill out forms and send them to Des Moines. The City of Dubuque had to get the approval of the state before it could be sent to Washington, D.C. When the federal government considered the grant for the Shot Tower, the city's developers had to let them know if they had or would need to raise the money to match the grant. Dubuque already had the money. After the money was collected from the grants and the America's River Project, the city had a grand total of $595,000. There were private donations that pushed it up to $600,000. The State Historical Society also contributed matching funds for this federal grant.

The success of this restoration project depended on the cooperation of many different people, civic groups and government agencies working together as a team. Many months of preparation and mountains of paperwork had to be conquered. The city is still leading this huge task, and with the talents of the Chicago restoration firm of HPZS and the experienced laborers necessary to achieve the actual work, the Shot Tower will once again stand as mighty as the river that flows by it. The plans for the project are now complete (2005), and the Shot Tower will be showing an updated appearance soon. Let the restoration begin!

Chapter 35

A 2001 Structural Analysis of the Shot Tower

By Ron Balmer

The City of Dubuque retained IIW Engineers & Surveyors, PC, in 2001 to evaluate the structural integrity of the Shot Tower. Every step of this assignment led to a deeper inquiry into the tower's history. In the end, we accumulated about four times as many images and exhibits related to the Shot Tower as were ultimately used in our one-hour presentation tracing the history of the tower from its inception to the present day. We showed this visual production to the Central students twice, and as I recall we never managed to get further into the Shot Tower's history than the 1911 fire within the allotted times. The Central classes were a joy, with students asking excellent questions and stimulating genuine discussion.

IIW began its 2001 investigation with an old linen blueprint of the first plat of "Du Buque" dated 1835 that resides in my office. By lining up and extending streets, it locates the Shot Tower beyond the islands and sloughs of the Mississippi River shown on that document. The tower is clearly described in the reports of the day as 26 feet square at the base with 9-foot-thick walls, 110 feet of stone and 40 feet of brick. IIW measured the base to be 19 feet square with 3-foot-thick walls, 80 feet of stone and 40 feet of brick as it stands today. The proportions of the tapering structure are consistent with stone extending below the current adjacent grade to a depth of about 30 feet, which would result in a footprint of about 26 feet. This implies the capability to build the tower foundation either on an island in the river or in the river itself, about 20 feet below the nominal water level, before the importation of fill raised the grade around it by about 10 feet in 1858.

The accurately portrayed circa 1857 woodcut image depicts the lowest tower wall openings about ten feet above grade and an attached wood shack on stilts. A 1930s-era drawing illustrated stone masonry stepping out at the current base elevation adjacent to the lowest wall openings and other features that are not consistent with historical information and early depictions. Fundamental to our investigation of the structure's integrity was determining exactly what the original configuration was relative to what we were looking at. The mythology surrounding the Shot Tower and the complexity of its history were fascinating and led us to the Dubuque County Courthouse, the Dubuque County Historical Society and the Center for Dubuque History at Loras College, as well as numerous other sources.

IIW's physical examination of the tower revealed considerable surface deterioration. However, the mass of the stone and brick walls remained largely intact, and there was no evidence of settlement or structural cracking. It became apparent in the course of our observations that the 1961 preservation project, called Save Our Shot Tower (SOS), though well meaning and executed consistent with standards of the day, had been marginally effective in protecting the structure. It was also apparent that no further stabilization, maintenance or restoration work had been performed on the structure since that time.

Prior to this 1961 preservation project, nothing had been done to the Shot Tower since the Standard Lumber fire of 1911. The top of the structure was now open to the elements, and considerable erosion of the soft brick and mortar was evident. In 1961, the brick was repaired and a flat pre-cast concrete slab was placed atop the tower for a roof, which was equipped with a hatch opening. However, no roofing material was applied to it, and as a result, water migrated through the porous concrete and attacked the repaired brick mortar joints immediately below. The wood-framed windows of the tower had been destroyed in the 1911 fire, and the masonry openings were left to deteriorate until wood louvers with bird screens were installed in 1961. By 2001, numerous louvers were broken and many screens were compromised, such that birds roosted inside at will. As a result, we encountered a pile of bird droppings several feet deep inside the tower. The soot from the 1911 fire still stained the interior. However, the masonry itself was in reasonably good condition.

On the exterior, one of the most striking observations was the hard mortar shells that were suspended out in front of the old lime mortar joints. Weather exposure had literally eroded the old lime mortar out from underneath the harder, 1961 re-pointing mortar that had essentially been troweled over the relatively wide stone mortar joints. The 1961 mortar mix had likely been formulated with a high Portland cement proportion with the best of intentions to provide greater strength and speed production. However, current preservation techniques call for a high lime content mortar to make it sacrificial to the masonry units. The original structure was built before the development of Portland cement, so the thick masonry walls had been laid with pure lime mortar, which requires more frequent maintenance when exposed to weather and preserves the integrity of the masonry units themselves due to its cushioning effects.

Lime mortar also requires a significant moist curing period to support continued vertical construction. However, it is very forgiving to structural movements when fully cured, conforming to sometimes significant deformations without cracking. Our research uncovered clues that the construction of the Shot Tower actually began in 1854, and we theorize that the top of the stone construction was reached in early 1856, indicating the relatively slow rate of production, primarily due to the limitations of lime mortar when setting large stone units that appear to have been quarried in northern Illinois. The February 1856 mortgage of the tower to finance completion likely signaled the change to brick masonry, which was probably manufactured in beehive kilns near Lancaster, Wisconsin, to reduce cost and speed construction, which was completed in November 1856. We observed numerous heavily fired bricks that appeared to have replaced eroded units in 1961, as well as original soft brick that we believe birds had burrowed into over the years.

Considering the neglect and lack of maintenance over the life of this edifice, the Shot Tower's survival and fundamental integrity are a testimony to the forgiving and enduring nature of traditional masonry structures.

Chapter 36

Rehabilitation of the Historic Dubuque Shot Tower

By Laura Carstens, Planning Services Manager, City of Dubuque

The Dubuque Shot Tower—a local, state and national treasure—is listed in the National Register of Historic Places and owned by the City of Dubuque. Constructed in 1856 to manufacture lead shot ammunition, the Shot Tower is one of only a few remaining shot towers in the United States and is the only one west of the Mississippi River.

Rehabilitation

In 2001, the City of Dubuque determined that rehabilitation of this historic landmark was in order and had a structural assessment made of the Shot Tower. The assessment revealed many areas of damage and deterioration that needed attention. The city committed $320,000 for the rehabilitation and interpretation of the Shot Tower as part of the America's River project at the Port of Dubuque. Grants from federal, state and private sources were pursued for several years before the city was successful in garnering enough matching funds to initiate rehabilitation.

Originally, the Shot Tower Rehabilitation Project was to complete the structural repairs of tuckpointing and installing new windowsills, windows and a new door. Repairs were to be done on the barricade walls and the electrical work, as well. Plans also called for capping the Shot Tower with a flat roof and constructing an interior stairway to the top of the tower. However, due to the high costs of building the stairway and the severe deterioration of the interior brick surfaces, this part of the rehabilitation had to be phased in.

Phase I, completed in 2010, involved the following: an archaeological survey inside and around the tower; repair, re-tuckpointing and restoration of the interior and exterior limestone and brick masonry surfaces; removal and replacement of the existing concrete roof cap with a new concrete cap and vent; and installation of new wood windows, louvers and door. Phase I provided an opportunity to interpret the Shot Tower's rich history, the Mississippi River and the country's industrial expansion westward. During the archaeological investigation, the well where lead shot was once dropped and cooled was unexpectedly uncovered. The well will remain exposed and become part of the interpretive experience.

Phase II will involve adding a roof hatch and the installation of a full-height, steel interior stair and handrails. The 1911 fire destroyed the original wooden stairway, leaving no ready means to access the tower's interior for maintenance and repairs.

A weather-beaten and vandalized Shot Tower window before Phase I of the most recent rehabilitation began. *John Adelmann.*

Phase III will require substantial site work around the Shot Tower's base to reopen the original doorway on the north side of the structure. An enclosed observation deck will also be added to the rooftop.

As of 2011, Phases II and Phase III currently are being deferred until a future date when additional funding can be secured. As a result, the flagpole that Central Alternative High School students raised money for through sales of the Shot Tower calendar also is being deferred.

REHABILITATION TEAM

Consulting Engineer/Architect: Hasbrouck Peterson Zimoch Sirirattumrong (HPZS) of Chicago, Illinois
General Contractor: Klauer Construction Company of Dubuque, Iowa
Subcontractors: Fred Jackson Tuckpointing, Morse Electric and Windco Roofing of Dubuque, Iowa

RESEARCH, MITIGATION AND ADMINISTRATION TEAM

Archaeological Survey Consultant: Great Lakes Archaeological Research Center (GLARC) of Milwaukee, Wisconsin
National Historic Landmark Application: History Pays! of Des Moines, Iowa
Hazardous Material Removal (bird guano): Discount Wildlife Service of Dahinda, Illinois, and Allied Waste Services of Dubuque, Iowa
Phase 1 Environmental Survey: Iowa Department of Natural Resources
Structural Assessment and Physical Survey: IIW Engineers and Surveyors of Dubuque, Iowa
Grant Administration: East Central Intergovernmental Association (ECIA) of Dubuque, Iowa
Historic Preservation Review and Compliance: National Park Service and State Historical Society of Iowa

PARTNERSHIPS

Rehabilitation of the historic Dubuque Shot Tower is a story of partnerships. The $630,000 Phase I rehabilitation and interpretation of Dubuque's historic Shot Tower was accomplished through the following local, state and federal partnerships:

- $295,000 Save America's Treasures (SAT) grant from the National Park Service to address the Shot Tower's structural problems and interpretation. This competitive matching-grant SAT program is one of the largest and most successful grant programs for the protection of our nation's endangered and irreplaceable cultural heritage. The City of Dubuque applied three times before receiving this grant. The Dubuque Shot Tower is featured on the NPS webpage for the SAT program at www.nps.gov/history/hps/treasures.
- $100,000 Historic Sites Preservation grant (HSPG) from the State Historical Society of Iowa (SHSI) for the Shot Tower's structural rehabilitation. The HSPG Program provides funds through a competitive matching-grant program to acquire, repair, rehabilitate and develop historic sites that preserve, interpret or promote Iowa's cultural heritage.
- The SHSI also provided a letter of support to upgrade the Shot Tower from state to national significance in the National Register of Historic Places to help it qualify for the SAT grant.
- $15,500 Historic Resource Development Program (HRDP) grant from the SHSI for an archaeological survey. The HRDP Program provides funds through a competitive matching-grant program to preserve, conserve, interpret, enhance and educate the public about the historical resources of Iowa.
- $320,000 from the City of Dubuque for the Shot Tower's rehabilitation and historic interpretation as part of the local commitment for America's River, a $188 million riverfront redevelopment project at the Port of Dubuque.
- $5,000 from the Old House Enthusiasts of Dubuque, Iowa, for a 2001 structural assessment of the Shot Tower using a crane service.
- Teachers and students at Central Alternative High School collaborated with the city's rehabilitation efforts to research and document the history of the Shot Tower and then made this information available to the general public. Student essays about their Shot Tower research were published in the local newspaper. Students created a calendar that commemorated the Shot Tower's

150th anniversary, the sales of which raised $1,185. The students' work enhanced the Shot Tower's educational and interpretive value by creating a permanent printed historical record of this unique and nationally significant historic structure. The project also benefited the students, enabling them to work directly with city government, to develop research and documentation skills and to learn about and appreciate the history of their local community.

- Jim Jacobsen, architectural historian with History Pays!, donated services for a National Historic Landmark (NHL) application to the National Park Service. Work on the NHL application was deferred pending completion of the rehabilitation at the request of the NPS.
- Loras College students in the Archeology and Cultural Heritage Interpretation Program participated in the archaeological survey conducted by Great Lakes Archaeological Research Center. Students were able to gain valuable hands-on experience in an archaeological investigation through this project, and the city's cost of the archaeological survey was minimized.
- The Iowa Department of Natural Resources conducted a Phase II environmental survey for possible residual environmental contaminants from the historic lead shot production at no cost to the city. This survey will supplement the NHL application for the Shot Tower.

Chapter 37

Hasbrouck, Peterson, Zimoch, Sirirattumrong (HPZS)

By Stacy Mootz

In March 2005, the City of Dubuque reviewed proposals from five different firms, searching for registered engineers or architects who were experienced in preserving, rehabilitating and conserving historic structures such as the Dubuque Shot Tower. These companies would also be expected to prepare plans for how they would complete the work. The firms that submitted proposals were as follows: Hasbrouck, Peterson, Zimoch, Sirirattumrong (HPZS) from Chicago, Illinois; Atkinson-Noland Associates/ The Collaborative from Boulder, Colorado; Steven Ulsted Architects, IIW Engineers and Surveyors, Dubuque, Iowa; and Anderson Design and Consulting of Dubuque, Iowa.

After reviewing each proposal, the city's consultant selection committee recommended HPZS because of their large amount of experience on similar projects involving historic structures with work done according to the Standards for the Treatment of Historic Properties. Consultants from Boulder, Colorado, listed projects they completed that met those standards also, but their cost estimate was over budget. HPZS has also completed rehabilitation and restoration work using the historic standards on many old but significant lighthouses, which are similar in design to the Dubuque Shot Tower. HPZS was the only firm to list this kind of experience.

Of the $600,000 total cost for repairs, the city set aside $60,000 for architectural and engineering services. HPZS estimated their costs would be $57,400, which was mid-range of the five proposals. On June 6, 2005, HPZS was officially approved to take on the renovation of the Dubuque Shot Tower. One member of the HPZS team is Mr. James Peterson. I contacted him with some questions I had about the Shot Tower rehab project and how his company would go about restoring it. He explained why he was interested in taking on the rehabilitation process, saying, "I have found the Dubuque Shot Tower interesting for some time, and since my wife went to college in Dubuque, we would talk about it whenever we visited Dubuque and always hoped that it could be saved as an important part of Dubuque's history and the history of lead shot, as well as remain as a visual landmark."

The rehabilitation of the Shot Tower in Dubuque will include stabilizing the brick to keep it from shifting and replacing the limestone sills and damaged brick/limestone units to stop birds from entering the structure. It will also include a fireproof set of metal stairs inside the tower that will reach all the way to the rooftop to help improve the tower's maintenance and stability. In addition, the plan calls for replacing the old pre-cast concrete cap with a new roof to prevent further deterioration of bricks inside the structure. Along with the stairs to the top of the tower, there are plans to build an observation deck, which will make it possible for citizens to get unique views of the city and the surrounding area. The City of Dubuque and HPZS are also planning to install a sturdy wall to protect the tower from damage from railroad cars and semi-tractor trailers that serve the industrial site next to the Shot Tower.

Peterson says the hardest part of this project and of most other historic rehabilitation projects is that often the building being worked on is no longer operating in ways that it became famous for. Because of this, sometimes it's difficult to convince people to go ahead with the restoration. When it comes to the Shot Tower, Peterson said it will be important "to find a continuing use for a structure that is no longer needed to make shot. I believe that the renovated and restored Shot Tower can continue to serve Dubuque as an important component of the renovated River Walk, as an observation tower and a visual focal point."

HPZS has been providing architectural and engineering services to clients through its offices in Chicago, Illinois, since 1976 and has received numerous awards for their work. Mr. Peterson told me, "Our firm has focused much of its practice fields of architecture and engineering on historic restoration of landmark buildings and structures. That focus has provided us with the opportunity to work on many local, national and international landmarks."

Their architectural and engineering services include programming and analysis, design, contract documents and site supervision, interior design, adaptive reuse of materials and conservation analysis and research. Some of the projects they have completed include the Carysfort Reef Lighthouse in Florida, the Dana Thomas House in Springfield, Illinois, the Seneca Grain Elevator in Illinois and various other structures around the United States.

In addition to the $60,000 consultant fee, it will cost $13,000 for the archaeologist, $7,000 for removing hazardous materials and $520,000 for the materials and labor that will be needed to complete the rehabilitation. The total cost of the project is $600,000. The process of working on the Shot Tower is estimated to take 182 days. As for the completion date for the rehabilitation, Jim Peterson said, "When we started our work on this project, we were looking at a schedule geared toward completing the rehabilitation construction activity this year [2005]. At the present time, it would appear as though the schedule has been pushed back a year."

Much like the newspaper reporter who went inside the Shot Tower in 1856, one day in the near future, Dubuque citizens can expect to enter the Shot Tower, see up close how it was built, walk up the stairs inside and learn firsthand of its important history.

Editor's note: When Stacy wrote this chapter in 2005, she took information from the proposed rehabilitation plans that both the City of Dubuque and HZPS had agreed upon. As the project got underway, modifications to the original plans were made. As of 2010, there will be no interior stairway and no observation deck on top of the Shot Tower.

Chapter 38

The Shot Tower Class Changes Direction

"Focus on Youth" essay by Nick Sprague

One day, I came to school with a cool idea—to create a calendar commemorating the 150th anniversary of the Dubuque Shot Tower. No other student or teacher thought about this. I got the idea thanks to my stepdad. He brought home a calendar that celebrated the 150th anniversary of the Morrison Brothers Co. of Dubuque, where he works. It is interesting that the Morrison Brothers company is located one block from the Shot Tower. Their calendar got me thinking of celebrating the anniversary of the Shot Tower in the same way. We had a discussion in class, and the more we talked, the more the calendar sounded like a great idea. Even though we started out with the goal of creating a Shot Tower research book, we decided to take the information we were discovering and put it in the form of a calendar that people could enjoy for a whole year.

We started looking for pictures and information we could use. But this history project would have been impossible to create without the help of some dedicated and talented citizens. My teacher, John, contacted graphic designer Joe Deppe, who agreed to help us with the design and layout of the calendar. Not long after talking to Joe, we found out there were other people in the community who wanted to get in on developing the calendar.

One person was Mike Gibson, the director of the Center for Dubuque History at Loras College. He helped us gather photos, old newspaper articles from microfilm and other significant historical information.

The George W. Rogers & Co. Shot Tower commemorative calendar, winner of the State Historical Society of Iowa's Loren Horton Community History Award and the Iowa Historical Preservation Alliance's Preservation at Its Best Award, both presented in 2007.

Tacie Campbell is the curator of the Dubuque County Historical Society (2006). She also has helped us find rare images and important material that we can use. Greg May, of One Step Printing, is going to help us print and distribute our calendar. In addition, one of my classmates, Jay Derinski, will be creating a detailed drawing of the Shot Tower that will go on the calendar's cover.

I am really excited that my idea, to honor the Shot Tower with a special calendar, is being supported by so many people. For years, the Shot Tower has sat by the Mississippi all alone and forgotten. It has survived many years of wear and tear, and it suffered through a huge fire and a direct hit from a runaway railroad car. Businesses have been built around it, and some people actually thought it should be moved to the Ice Harbor, but the Shot Tower still stands proud and tall.

Today, Central Alternative High School students are taking the task of preserving a piece of our city's history seriously. With the help of the community, we will bring the Shot Tower the respect it deserves. I am very proud and thankful that my idea to honor the Shot Tower with a calendar will soon become a reality.

Chapter 39

Drawing the Shot Tower

"Focus on Youth" essay by Jay Derinski

As a student in Central's Shot Tower research class, I was asked to complete a detailed drawing for the Shot Tower calendar showing how it appeared 150 years ago. I did this by studying an old woodcut image that was done in the 1850s to get an idea of how it looked. I had no problem doing this because I like to draw. I used a very fine black ink pen to draw the individual bricks and limestone blocks, and all the windows and their frames, and tried my best to make it as accurate as possible.

One of the major details of how the Shot Tower used to look is the wood outbuilding that was built next to it. From the research we have done, we believe this structure was used for storing sacks of finished shot and other shot-making equipment until its floor collapsed in 1862. I made sure to carefully draw the wood planks with the different grain patterns to make the image realistic.

Another major feature of the Shot Tower is its many windows. The airflow through them could have interfered with the shot-making process by blowing the melted lead all over the walls instead of falling straight down to the tub of water at the base to cool. If you look at other shot tower structures across the country that were built many years ago, you will find that there aren't many windows in them, especially in the shafts where the shot is dropped. Some historians don't understand why the Dubuque Shot Tower had so many windows.

In addition to having his artwork chosen for the cover of the Shot Tower calendar, Central Shot Tower Investigator Jay Derinski's pen-and-ink drawing was featured on the cover of *Julien's Journal* in August 2006. *Julien's Journal.*

Just before school got out for the summer (2006), we were contacted by the local news magazine of Dubuque, *Julien's Journal*, after our calendar came out. The editors asked if it would be possible for them to use my Shot Tower pen-and-ink drawing for the cover of their magazine. I was honored that they made this decision. Central's Shot Tower research project has given me the chance to display my artwork for the entire community to see. It was great to be able to tell people about what we have learned in class about Dubuque's famous structure.

The Shot Tower researchers at Central Alternative High School are very proud to be telling the entire community about the history of this grand old building. It is an important story that will help us understand how life was in Dubuque over 150 years ago.

Chapter 40
2006: The Year of the Shot Tower

"Focus on Youth" essay by Brian Mess

In my very first day at Central Alternative High School, something happened to me before I could even sit down in the Shot Tower class: I was asked if I was going to attend the Dubuque City Council meeting later that night. They were going to read a proclamation saying that 2006 was going to be the year of the Shot Tower. Why is this Shot Tower so important? Why is it such a big deal to do this anyway?

The main reason these issues are so important is because the Shot Tower has been standing tall for 150 years since it was first opened for business on December 24, 1856. Also, our Shot Tower was added to the National Register of Historic Places in 1976 for having national significance for making and distributing lead shot during the 1850s and '60s, and it's the only shot tower west of the Mississippi that is still standing. Hopefully, this spring (2006) for the Shot Tower's 150th birthday, the City of Dubuque will undertake a complete building rehabilitation.

This will be the first repair work done to the Shot Tower since 1991, when a runaway railroad car bashed into the west side of the tower and caused $10,000 damage. The Shot Tower renovation also is part of the development of the Mississippi Riverfront and River Walk. So far, the City of Dubuque has raised $595,000 to help pay for the work that is going to be done on the structure. Finally, the Shot Tower is getting the attention it deserves after all these years of sitting abandoned.

When we got to the city council meeting, all the chairs for the members were in a row with microphones and computers. As the meeting's agenda got underway, several people were called forward to receive proclamations that Dubuque mayor Roy Buol read aloud and signed for their groups. Then it was our turn—the students from Central. Mayor Buol read the proclamation

DUBUQUE

PROCLAMATION

WHEREAS, the Dubuque Shot Tower was constructed in 1856 to manufacture lead shot ammunition and will celebrate its 150th Anniversary in 2006; and

WHEREAS, the Shot Tower is listed on the National Register of Historic Places as having national significance; and

WHEREAS, the Shot Tower is one of only a few remaining shot towers in the United States, and the only one west of the Mississippi River; and

WHEREAS, the Shot Tower is part of the $188 million *America's River* project of riverfront recreational and cultural venues, located at the Port of Dubuque; and

WHEREAS the City of Dubuque has been awarded a federal Save America's Treasures grant and a state Historic Sites Preservation grant for rehabilitation and interpretation of the historic Shot Tower; and

WHEREAS, this project will create an opportunity to interpret the Shot Tower, the Mississippi River, and the country's industrial expansion westward by providing access to this unique historic structure; and

WHEREAS, the City of Dubuque, in partnership with the Convention and Visitors Bureau and Central Alternative High School, will celebrate the restoration of the historic Dubuque Shot Tower through various community events during 2006.

NOW THEREFORE, I, ROY D. BUOL, MAYOR OF THE CITY OF DUBUQUE, IOWA DO HEREBY PROCLAIM 2006 AS THE

"YEAR OF THE SHOT TOWER"

IN DUBUQUE, AND CALL UPON THE PEOPLE OF THE DUBUQUE AREA TO SUPPORT THE RESTORATION OF THIS UNIQUE AND NATIONALLY SIGNIFICANT HISTORIC STRUCTURE IN 2006.

IN WITNESS THEREOF, I have hereunto set my hand and caused the Seal of the City of Dubuque, IA to be affixed this 17th day of January, 2006.

Roy D. Buol, Mayor
Attest:

Jeanne F. Schneider, City Clerk

A proclamation: "The City of Dubuque, in partnership with the Convention and Visitors Bureau and Central Alternative High School, will celebrate the restoration of the historic Shot Tower through various community events in 2006." *Center for Dubuque History.*

declaring 2006 as the Year of the Shot Tower, signed his name to it and presented it to several students in the audience who were asked to come up and receive it. Then we gave him two special gifts. One was a certificate stating he was an honorary member of our Shot Tower research class, and then we presented him with one of our CSI T-shirts. CSI stands for "Central Shot Tower Investigators." Central student Spencer David then went to the podium and spoke to the council for all of us, and he got in a pretty funny joke when he said that we were glad to present Mayor Buol with this T-shirt but if anyone else on the council wanted one, they would have to buy their own! Everyone at the meeting laughed out loud.

When the Shot Tower is rehabilitated, it will provide a perfect opportunity for Dubuquers and visitors to our town to learn more about this building and appreciate the country's industrial expansion westward. It will also give people the chance to understand why this unique historic structure was so important during our city's early history. I am also very proud to say that the city's proclamation states in part, "The city of Dubuque, in partnership with the Convention and Visitors' Bureau, and Central Alternative High School, will celebrate the restoration of the historic Dubuque Shot Tower through various community events during 2006." My classmates and I are ready to take this challenge to the next level to find and reveal the truth about Dubuque's famous Shot Tower. Wow! What a class!

Chapter 41

The Archaeology and Geomorphology of Site 13DB838

By Michael M. Gregory, PhD

Archaeologists and geomorphologists from Great Lakes Archaeological Research Services, Inc., and its subcontractor, Strata Morph Geoexploration, Inc., investigated the George W. Rogers & Company Shot Tower (site 13DB838) during 2006 and 2007 in pursuit of answers to research questions for which documents are either silent or provide conflicting accounts. As a result, the research focused upon determining (1) how deeply the tower's base extended below the modern ground surface and whether this data may resolve conflicting accounts about its original height; (2) if pilings, cribbing or a combination of techniques support the structure; (3) the nature of the deposits filling the tower's interior; (4) the type of catch basin into which falling shot collected and if the basin lay exposed during 1934 when the Historic American Buildings Survey (HABS) documented the structure; (5) the presence or absence of lead shot in lower fill deposits; and (6) if undocumented architectural features exist. The investigation concluded with mixed results.

The study area that surrounds the Shot Tower has maximum dimensions of 125 feet east–west by 82 feet north–south and encompasses about 0.24 acres. Within the project area, the Shot Tower takes up about 19 feet square and is surrounded by a low stone wall supporting a wrought-iron fence. In addition, the fenced area is composed of blacktop and grassy areas, which together define approximately 0.02 acres. Due to the presence of this

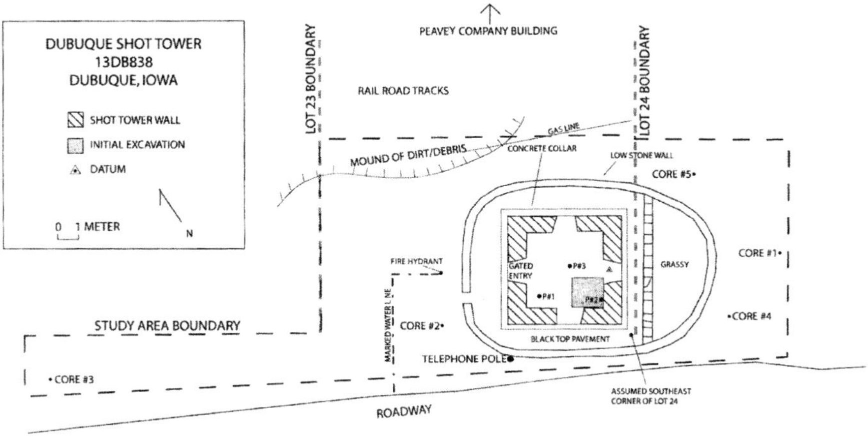

The plan view developed by archaeologist Michael Gregory showing the 2006–7 study area of the Shot Tower's footprint and the locations from where ground core samples were taken to determine the nature of the structure's subterranean foundation. *Michael Gregory.*

barrier, researchers could not investigate the area immediately adjoining the Shot Tower.

Restricted by structural features and utility lines, the geomorphological investigation resulted in the mechanical extraction of five soil core samples from outside of the fenced area and the manual collection of three soil probes from within the tower. Analysis of the soil cores extracted immediately outside of the walled area identifies four stratigraphic sequences: asphalt/fill, dredge spoil, wood/fill and alluvium. Of these, the wood/fill sequence is the most important regarding the tower's construction.

The top of the wood/fill sequence is marked by a weakly developed soil horizon formed in silty and sandy deposits and may be similar to the ground surface present at the time of the tower's construction. This surface begins 6.9 feet to 13.0 feet below the modern surface and rests upon beds of sand interstratified with thick, solid pieces of wood—possibly lumber, cribbing or railroad ties. The recovery of clinkers (partially melted lead ore mixed with ash), cinder and wood fragments from the upper soil horizon indicates the deposit is historic in age and is most likely the landscape surface disturbed by the construction of the tower. Beneath this sequence is a sandy and silty, wood-free alluvium that predates the tower.

Evidence of thick, solid pieces of wood are present in the three core samples taken nearest the Shot Tower, while only wood fragments occur in the cores extracted farther from it. This suggests that the area immediately around and beneath the tower was excavated prior to construction to accommodate the placement of the catchment well, wood cribbing, shoring or scaffolding. The smell of creosote also emanated from several core samples, but whether this is from the wood or the surrounding sediment remains uncertain. Across the United States, the widespread treatment of railroad ties with creosote did not begin until well after the time of the tower's construction in 1856; whether creosote served other purposes related to the construction or use of the tower remains unknown.

The archaeological investigation yielded new data about the structure and its history. Initial excavation within the southeastern corner of the tower exposed a segment of the feature's structural base, including a portion of the stepped foundation, and revealed that these structural elements lay exposed during the 1934 HABS study. In addition, a feature exposed but not previously recorded is a series of carved rock floor joist supports and a narrow rock ledge used to support the wooden floor or landing of the tower's lower level. The joist supports rest on the upper surface of the stepped foundation.

Encouraged by the discovery of these undocumented architectural features, the excavation proceeded with the removal of all interior fill. In a

Narrow slats were cut into the base of the Shot Tower's interior limestone and masonry construction by the builders to accept wooden floor joists. These previously unknown and undocumented architectural features were discovered during the 2007 excavation. *Michael Gregory.*

herculean effort, archaeologists removed soil from the interior area measuring about 13.0 feet by 13.0 feet to a depth of about 10.0 to 13.5 feet below the entryway sill and, in the process, achieved a number of research goals. Excavation exposed and removed all dredge spoil covering the circa 1940 surface, which closely corresponds to the surface documented by the 1934 HABS study and, where extended below the surface and through underlying fill, revealed the top of the 1911 fire debris layer. The existing water table prevented excavation of the debris layer, which was manually probed with a soil corer, revealing deposits extending at least another 2.92 feet deeper where the probe did not encounter buried, presumed wood beams. In addition, the excavation revealed the previously documented interior, tiered foundation in its entirety, as well as the original north wall entryway and three sets of previously undocumented floor joist supports in each corner, similar to that exposed in the southeastern corner. Unfortunately, the water table prevented archaeologists from excavating below the top of the 1911 fire debris to address all the research questions. The lead shot catchment well basin could not be exposed, demonstrating that while HABS workers accurately documented parts of the interior foundation, they did not observe

A rare look inside the Shot Tower. Loras College archaeology professor Cindy Smith excavates the eastern interior corner at the 1940 ground level. Deeper excavation at left reveals the 1911 debris level. Floor joist grooves are also visible. *Michael Gregory.*

the well basin, which is now known to be conjecture on the scaled HABS sketch. Similarly, the work did not extend to a depth where lead shot, if present, occurs. Finally, the investigation did not produce data about the depth or construction of the foundation. This is disappointing because these data can be used to resolve the debate about the tower's finished height. Earliest accounts state the tower was to be 150 feet high, but this contrasts with later accounts describing the tower as 130 to 190 feet tall. A careful evaluation of the documentary data suggests the Shot Tower stood between 130 feet and 150 feet high, and the difference between these two figures can probably be accounted for by the reference points associated with a measurement, whether measured from ground surface to rooftop or from foundation base to rooftop.

The subsurface investigation of the Shot Tower yielded a large body of archaeological, geomorphological, historical and architectural data that contributed to a more complete understanding of the tower's construction, use and documented history. In addition, the information provides guidance for the future drafting of a more detailed and accurate restoration design plan for the Shot Tower. While many of the research goals set for the project have been achieved, others were not, but in the research process, archaeologists and geomorphologists gained valuable insights as to how future investigations may successfully approach the issues and resolve existing debates about the Shot Tower's construction, height and internal layout.

Chapter 42

An Interview with Mike Gibson

"Focus on Youth" essay by Kristi Morrison and Andy Sawvell

When it comes to looking up information on the Dubuque Shot Tower, you won't be able to find it in typical high school textbooks. You have to go elsewhere and dig deep, like at the Center for Dubuque History at Loras College. Our class traveled there to add to our knowledge of this famous local building. Many people are unaware of the impact the Shot Tower has had on the city's history, but students at Central Alternative High School want to change that.

"The Shot Tower is not just a place for pizza and beer!" said Mike Gibson, the center's director. He attended a luncheon we hosted for members of the community who will help us on the project, and we sat down and talked with him. We have already learned that in 1976, the Shot Tower was added to the National Register of Historic Places and that it is one of only seven shot towers still standing in the entire country. However, Mike sadly commented, "Our Shot Tower is the most neglected and least publicized icon in Dubuque. It needs some attention to heighten the community's awareness of it."

For the past twenty-two years, Mike has been researching the Shot Tower but has not witnessed what he called a "concerted effort" to learn more about it. He said, "The only research that has been done on the Shot Tower by students has been for an occasional speech or English class. No students have gone as in-depth as those at Central are planning to do." Gibson added, "This will be a great opportunity for the younger generation to shed light on the Shot Tower and to inform the public of its significance in our city's history."

Mike's enthusiasm is motivating Central students to make a difference in their community. We are thrilled at the chance to dig up information about our city's past and look forward to contributing to the future. We are also privileged to visit and study in the Loras archives with Mike. He hopes that our efforts will educate the community and dispel any myths or false information that have become part of the Shot Tower story to this point.

Like we said, finding information on the Shot Tower is going to be challenging. We are determined to uncover the hidden truth. The Shot Tower has survived 150 years of our country's and our city's history. We will help to reveal, honor and preserve it.

Chapter 43

Discovering History in Our Own Backyard: The Key to Appreciating the Past

By Michael D. Gibson, Director, Center for Dubuque History, Loras College

In the 1970s, while working for the State Historical Society of Iowa in Des Moines, I became impressed with a resource kit that was being created for Iowa's elementary schools. At that time, Iowa history was taught at the fifth-grade level, and this kit contained classroom materials that illustrated how events in our state's past related to our national history. When I moved to Dubuque and became immersed in its rich history, it occurred to me that a similar resource kit could be developed that would focus on the story of our city.

In 1989, with a grant from the Dubuque Racing Association, the Center for Dubuque History at Loras College partnered with a group of active and retired elementary teachers, many of whom were interested in social studies. After much hard work, sixty-eight resource kits were produced and distributed to third-grade classrooms throughout the county and became a formal part of the Dubuque Community School District's third-grade curriculum. They are still in use today. Stocked with copies of historical documents, books, photographs, oral histories, art, games, ideas for activities and field trips and music composed by local teacher/historian/musician Lou Fautsch, these kits became valuable tools that stimulated students to become excited about their local history. During the initial in-service sessions, teachers were told that these kits were meant to be starting points, the contents of which could be utilized according to what worked best with each new class. They were not to sit on shelves and gather dust but rather be adapted to meet the constantly changing needs of both students and teachers. Their purpose was to encourage young people to take an interest in the history that surrounds their everyday lives—the history of their own backyard.

When these resource kits were introduced, they replaced units on Native Americans of the American Southwest. However, I had to convince skeptical teachers and administrators that the pupils were not being short-changed, because now they would have the opportunity to discover the Native Americans who lived right here in the Dubuque area. Wouldn't the students be more interested in the stories of local people than of those who lived many miles away? Is it not more relevant and meaningful for Dubuque students to learn about local lead miners, local entrepreneurs and local pioneers of over a century ago who helped build the town and state we live in?

About the time these local history resource kits were receiving acclaim in local elementary schools, I was approached by two teachers from Central Alternative High School in Dubuque—John Adelmann and Tim Ebeling. For several years, they had been working with their students, researching and writing books on various historical topics. But these teachers also wanted opportunities to conduct in-depth research into areas of local interest. Obviously, the Center for Dubuque History's rich collections intrigued them. Soon, another rewarding partnership developed, and Central Alternative High School students became regular patrons of the center for a span of more than twenty years. Through the heroic efforts of John and Tim, and countless hours of research by hundreds of their students, local history became a viable part of their school's educational experience. These efforts paid great dividends as the young people from Central, many of whom had not experienced much academic success in traditional school settings, researched history, published books and put on public seminars that made local connections with many notable Americans from history, including the Tuskegee Airmen, the civil rights movement's "Little Rock Nine," Mercury astronaut Scott Carpenter and the man who planned the atomic strike mission to Japan during World War II, Brigadier General Paul Tibbets. They succeeded in developing personal relationships with people and events from the past they would never forget.

In 2005, the students in Adelmann's and Ebeling's classes were given the challenge to conduct an in-depth study of the historic Dubuque Shot Tower. It had always been my hope that citizens in Dubuque would come to know the difference between our two Shot Towers—one being the historic landmark down by the Mississippi where lead shot was produced in the 1850s and 1860s and the other being that popular downtown restaurant where people enjoy pizza and beer. Up to this point, nobody had really made an effort to scour old newspapers and other primary sources to learn about our city's famous icon. Once the students began their research, they also discovered that nobody had pulled together the various images of the Shot Tower. As a result, they decided that a pictorial wall calendar featuring factoids of Shot Tower information would help fill this void. Their timing was perfect, as 2006 marked the 150-year anniversary of the construction of the Shot Tower. In order for everything to work, the students had to sell their idea to businesses and civic groups to fund the project. The community supported them, and the students stepped up to help design, promote and sell their calendar. In the process, they not only developed into local Shot Tower experts, but they also became determined to share with others what they learned because they had become inspired by their local history.

I have always been aware that history is often a tough sell, especially to young people. Today's youth seem to live only for the now, with little or no interest in yesterday or tomorrow. Many older people visit the center and tell me they wish they had taken more of an interest in their family's history years ago, but now it was too late. Had they done so, they could have made better sense of those boxes of old photographs in the attic, they could have interviewed grandparents and they could have learned about their ancestors' lives to discover more about their own. People ruefully say that they hated history when they took it in school because it was boring or irrelevant to them. It is for these reasons that I firmly believe we must find ways to get young people interested in the fascinating stories about individuals and events of the past. I am convinced that one way to do this is to allow them to learn something about the local history that exists in their midst. Local history can truly be the hook that grabs them and kindles their interest. The research the Central students did that inspired them to become ambassadors for the Shot Tower is a perfect case in point.

Hopefully, readers of this book will come to understand the value of studying local history and learn from the exciting discoveries that these young people from Central—call them history makers—uncovered in their own backyard. The study of local history is an exercise that is interesting, relevant and meaningful. But most of all, it stimulates self-worth by developing in each individual a sense of place and belonging in time and enables each of us to make personal connections with the events of days gone by and with the people who helped shape who we are today. Without this connection, history is reduced to an irrelevant, mind-numbing list of names and dates. Remember the old Native American proverb: "Who neglects the past, risks losing the vision."

Chapter 44
Dave Simon: Student Mentor

"Focus on Youth" essay by Spencer David

If there is anyone in our Shot Tower research class who deserves recognition, it would be Dave Simon. Dave is a volunteer/mentor who meets with us twice a week to help our class do research. He and I talked about his interest in the class and why he would want to work with us.

When I asked him about his connections to the class and how he got involved, Dave replied, "It was a roundabout journey, starting with a conversation with Kim Swift, Central Alternative High School's principal; working with the Sunny Crest Newsletter project; and then being asked to work with the Shot Tower project."

I asked Dave what his main interests were concerning the Shot Tower. He answered:

> *I saw it as a child growing up in this part of town. I hiked around it as an adolescent, and now I've come full circle as a retired citizen. There are so many interesting aspects to the tower, such as the prehistory of the Dubuque area, Native American activity, the contact period of European discovery, the subsequent uses of the tower after shot production ceased, restoration efforts and future plans. What interests me most, however, are the business plans, the start-ups, the minor successes and failures and the determined attempts to make shot production successful in our city.*

What does he hope our research project will accomplish? "Ideally, I would like to see the Shot Tower become the primary symbol of Dubuque. It should represent not only the efforts of our city's founders but also should be a unique icon, like the St. Louis arch or the Eiffel Tower in Paris. It's the appropriate size and shape for our community, and there is no other shot tower that looks like it today."

When discussing the areas he has been researching, Dave told me, "I have particular interest in finding the 'shot shaft' that Julius Graves cleverly used for shot production after the Shot Tower was sold in 1862. My research led me to the old Union Park area, just down the creek from my home. I have obtained images of the lead deposits and crevasses, which may lead to discovering the site."

Dave Simon has inspired my classmates and me with his determination not to give up even when trapped at a research dead end. We thank him for his efforts to make our Shot Tower research project a success.

Chapter 45

Mentoring At-Risk High School Students

By Dave Simon

I was a volunteer mentor at Central Alternative High School from 2005 until 2010. Its location on lower Bluff Street is significant for me because I grew up in this part of town. In my youth, the school now called Central Alternative High School was then called Franklin Elementary. I attended kindergarten in this facility and grades one through eight at St. Raphael School, just one block north. Living on West Fourth Street provided me with a daily view of the Shot Tower, which was due east of my childhood home, so this research project was a good fit for me.

During my time at Central, I developed a gentle approach with the students in order to put them at ease and allow them the freedom to dig into the resources available at school as well as other institutions. I wanted the students to know that there were adults in the community who cared about them and wanted them to do well in life. I noticed that most students were eager to forge ahead if they believed they were respected and could rely on the support and approval of the adults in their school environment. Imagine, then, how much more valuable community-wide support can be for students who are struggling to become successful people in life.

When the Shot Tower project began, most students didn't have a clue about its historical significance. Even my knowledge was limited to its Civil War myths. However, as the students discovered the challenges and risks that were associated with building and operating the tower, it became easier for them to relate, as if they now had something in common with people of the past—people who decided to take a risk to achieve success. I believe this notion drove home to our students that they had the potential in themselves to succeed by working hard to achieve their goals. I often expressed to them my pride in the Shot Tower and its symbolic value of our city's early years. Usually somewhat disaffected and uncaring, the students in the class soon enough began to care about the Shot Tower. Perhaps it was their first sense of ownership and a personal belonging that they felt as citizens of Dubuque.

Some of the students responded well to the help I gave them and were trusting of our relationship. I could tell this by their responses to my suggestions and by how well they followed through with the necessary research and writing their topics required. An indication of this trust was revealed in the unguarded conversations we had while shuttling back and forth from school to our fieldwork sites. Loras College's Center for Dubuque History, the Dubuque County Historical Society's archives and the Carnegie-Stout public library are impressive institutions that can quickly put a person in the mood for doing research. Most students were awed by these facilities and responded well to the opportunities made available to them.

As research progressed and the students' knowledge of the Shot Tower and its history increased, I could see a sense of ownership developing in them. At one point, this attitude got so strong that the students felt confident enough to propose another way to tell the Shot Tower's story—a way that neither of their teachers had considered. They wanted to present their Shot Tower findings in calendar form. This was certainly an awkward point in the project. While publishing a book would offer a solid way to showcase the students' research efforts, a rejection of their calendar idea could have been a blow to their initiative and been counterproductive by denying them the chance to act on their inspirations. After all, they were only following the example that Dubuquers over a century and a half ago set in order to build the Shot Tower—take risks to succeed!

As time progressed, so did the spirit of the students. They were growing more and more confident that their work would eventually yield a first-class product, be it a book or a calendar. The students with whom I worked did not sense that they would fail, even though it would have been easy for them at various stages to give up. From time to time, some of them went down blind alleys in their research, but they would almost always recover and continue searching in different directions. My advice was always to keep moving forward, wherever they were.

What was really a surprise was that during the project some of the student researchers discovered family ties to early Dubuque personalities. Their investigations provided further knowledge about their ancestors and gave them more reasons to be proud of their discoveries. Names such as Rhomberg, Sproule and Burden abounded throughout Dubuque in the mid-nineteenth-century records and writings. But the most surprising discovery was made by student Sean Bosveld, and he did it in an unusual way. Sean enrolled in the class well after it had begun, and many of the topics directly associated with the Shot Tower itself were already taken by other students. He decided to research and write about Julius Graves, the Shot Tower's last owner. It was a topic no one else had chosen. Then, sadly, Sean's grandfather died. While at the Linwood Cemetery grave site, Sean happened to look around at nearby gravestones, saw the name Julius Graves and later mentioned to his grandmother that he was doing research on that man because he was one of the owners of the Shot Tower. That's when Sean's grandmother told him that Graves was his great-great-grandfather! This news energized the class, put a real charge under Sean and ended up with him getting a feature story in the *Telegraph Herald*. The article, whose headline read "Student learns ancestor towers over city," put him in the permanent history of Dubuque.

The Shot Tower research project was a positive experience for the students in so many ways. They learned about local history. They learned how to research and write about their topics. They reported their findings to the community. But most importantly, they developed a greater sense of self-worth that, hopefully, will become for them the foundation onto which they can add future successes as they mature and go forward. I am very glad to have made a contribution to their lives.

Chapter 46

Research Has Its Ups and Downs

"Focus on Youth" essay by Amanda Sproule

Students at Central Alternative High School are still hard at work dealing with the joys and frustrations of researching Dubuque's historic Shot Tower. Our major problem is finding concise and pertinent information about the structure's history. It's not like we can go looking in a textbook to find the answers. When information is found, we almost celebrate in class. In the beginning, any new information was easy to find because I had little knowledge of the tower. My fellow researcher, Spencer David, has been in this class since the beginning, and he proves my point.

"When I first started this class, I was excited. It was much easier to find information then, because it was all new to me," he said. "But as my knowledge grew, so did my questions as I learned more about the Shot Tower. As we go into this year [2006], it gets harder to find information that I don't already have. I have so many more questions, and now I can't seem to find ways to answer them."

Digging for the historic information is the biggest frustration, but when you find what you're looking for, you feel as though a weight has been lifted off your shoulders. In this class, my mind is constantly thinking, dwelling, day in and day out, on what more I could do. However, despite the challenges, my classmates and I are determined to reach our goal: to publish a book on the history of the Dubuque Shot Tower.

History class should not be boring. Sure, history has its ups and downs, but it's supposed to be challenging. Our Shot Tower class is just that. We have no textbooks, no teachers' guides, no table of contents, no quizzes or final exams at the end of the semester—just our minds hard at work trying to discover what we can about the Shot Tower. Sometimes you can find us at the Center for Dubuque History at Loras College, searching through old documents or scrolling through microfilm, at the Bowell Library in the National Mississippi River Museum, at the Carnegie-Stout Library or at the Shot Tower site itself, taking measurements and photographs.

Sometimes I feel as though I am putting together a huge jigsaw puzzle in this Shot Tower research, but all I have pieced together at this point is the border. The whole scene is still incomplete, but as my classmates and I set more and more pieces in place from our research, we'll begin to see the entire picture of the Shot Tower's significance in our city's history, and so will you.

Chapter 47

Shot Tower Class Still Hard at Work

"Focus on Youth" essay by Maranda Duhme

Our Shot Tower class is still hard at work (2006). However, at certain times there are more "newcomers" than veterans in the room because some students have to take other courses to graduate. But don't get the wrong idea—we are still making progress. When I asked a new student, Corey Adams, what he thought about this class, he said, "It's a great opportunity to learn how to research historical events in our own community." As you may know by now, in the final stages of this project, we will publish a book on all the information we gathered while doing our research and share it with the entire community of Dubuque.

When I asked another student who has been in the class for a while what he thinks the book will be like, Nick Copeland said, "The book that will eventually be written will let the community know that the Shot Tower is not just a place where you go to eat pizza and drink beer, like Mike Gibson said. It's a historical part of the town we live in, and we should know about the history of our community."

Speaking of the book, the history of Nicholas Perrot, Julien Dubuque, the Native Americans they interacted with and the lead mines they worked in are topics that I wrote about in another essay. These are just samples of what the book about the Shot Tower will contain. It will cover the days before the white man came on the scene in the Midwest, and it will also explain the work that is now being planned to improve the Shot Tower. Our book will reveal the untold story of the famous structure that is over 150 years old and is still standing by the Mississippi River. Even though there is a decent amount of information about the Shot Tower that we found in local history encyclopedias at Loras College and the Mississippi River Museum, many people have been telling us that there has never been a book written just about the Shot Tower itself. I am very proud to say that our book will bc the first of its kind.

When I talk to people outside the class who have gotten to know about the project, some of them are a little confused about how we can have a class based on just one topic that doesn't get boring after a while. For an explanation, I sat down with our student teacher, Nick Doland from Loras College, and asked him why he thinks just studying about the Shot Tower keeps so much interest going in our class. Nick said, "As opposed to the typical or traditional way of learning, this class is in-depth and focused on one particular topic. The students are making personal connections that mean something to them, and that's what keeps everyone motivated to learn more. For this reason, the students are willing to take a more active role in their class and in their research."

As a student in this Shot Tower research class, I agree with Nick in saying that the traditional social studies or history lecture class can sometimes be a boring and difficult way of learning. There are teachers who skip around from subject to subject, which sometimes makes it hard for students to remember things, but not at Central. Our class is researching and gathering information about the Shot Tower so that everyone can learn how significant it is. This will be a book that will help people appreciate the history of our city and will reveal the real truth behind the making of the Shot Tower, how it operated and how it has managed to survive for over 150 years. When it comes out, we know you will enjoy it.

Chapter 48
Shot Tower On-site Rehabilitation, Phase I

By John Adelmann

In the summer of 2010, the Fred Jackson Tuckpointing Company of Dubuque tackled the process of repairing and stabilizing the exterior surfaces of the Shot Tower. Using a device called the Hydraulic Mast Climber, workers removed loose and degraded mortar between the bricks and limestone, re-tuckpointed the surfaces and replaced the deteriorating, porous cement roof with a wooden frame wrapped in a 60-mil rubber membrane.

As a result of this work, the Shot Tower's masonry was restored to protect the exterior as well as the interior of the structure. For visual and aesthetic purposes, the tower also had interior and exterior lighting installed to accentuate its texture and height. Another aspect of this restoration centered on making sure that the Shot Tower will continue to be stable for the future installation of an interior staircase that will rise to the roof. These improvements should protect the Shot Tower for fifty to sixty years.

According to company president Kelly Jackson, the same business under his father Fred's direction did minor work in the 1960s to help stabilize the Shot Tower, and it was exciting for him to return to the project some fifty years later to make more extensive improvements that will serve to maintain the structure as an enduring symbol of Dubuque. (The elder Jackson also erected a scaffolding inside the Shot Tower in 1976 to help Joseph Pickett Sr. in his quest to drop shot from there—see Chapter 30.) Said Jackson, "This was a great project—we had a very good working relationship with the City of Dubuque and the general contractor, Klauer Construction."

Above: The Shot Tower's replacement roof, complete with rubberized surface, ventilation port and lightning arresters. Eyebolts were used to hoist the roof into position in July 2010. *John Adelmann*.

Right: The Jackson Tuckpointing work crew used the Hydraulic Mast Climber to create a safe working platform around the Shot Tower, regardless of the height. *Fred Jackson Tuckpointing*.

He added:

> *This is important historic preservation. We were part of a rehabilitation that has succeeded in protecting a significant piece of our city's past for future generations to enjoy. A structure like this would be built today with*

Left: Another view of the Hydraulic Mast Climber. Workers replaced window lentils, ground away loose mortar and re-pointed the Shot Tower's brick and limestone. *John Adelmann.*

Below: "What a glorious sight was presented to our eyes. A scene exhilarating in the extreme and worth the toil of climbing the height of one hundred and fifty feet to enjoy it." *Fred Jackson Tuckpointing.*

steel and sheet metal, not brick, limestone and wood! The craftsmanship that made this Shot Tower possible in 1856 should be preserved so people can appreciate the unique design and construction skills of days gone by.

Appendix

Shot Tower History Timeline

1644	Nicholas Perrot born in France.
June 17, 1673	French explorers Marquette and Jolliet reach the Wisconsin River.
June 25, 1673	Marquette and Jolliet contact Illinois tribe on the Iowa River.
1685	Perrot arrives in upper Mississippi River country and builds forts, including one at Prairie du Chien, where Wisconsin and Mississippi Rivers converge.
1690	Miami Indian delegation asks Perrot to locate across from Catfish Creek to trade pelts. Perrot receives gift of ore from chief.
Aug. 25, 1700	Frenchman Pierre Charles Le Sueur travels north from Gulf of Mexico on the Mississippi, arrives at the Fever (Galena) River, which he calls the "River of the Mine."
1702	Le Sueur returns to France with reports of mines around Dubuque.
1703	First map of New France reveals lead mines at Catfish Creek.
1710	Maps published in London note mines at Dubuque.
Aug. 13, 1717	Nicholas Perrot dies in Canada.
Jan. 10, 1762	Julien Dubuque born in St. Pierre-les-Becquets, Canada.
1783	Dubuque arrives in Prairie du Chien.

Sep. 22, 1788	Dubuque signs contract with Fox tribe at Catfish Creek for permission to work—not own—their lead mines.
1796	Dubuque gets approval from Spanish governor general Baron De Carondclet for his claim to "Mines of Spain."
June 1, 1803	Mines of Spain comes under jurisdiction of the U.S. in the Louisiana Purchase.
Oct. 20, 1804	Dubuque conveys 7/16 of the Mines of Spain to merchant Auguste Chouteau to cover his debts.
Sep. 1, 1805	While visiting this area, explorer famed Zebulon Pike gets evasive answers from Julien Dubuque about his mining activity and ore yields.
Mar. 24, 1810	Julien Dubuque dies and is buried on bluff overlooking Catfish Creek.
1824	James Langworthy arrives in Galena, Illinois, from St. Louis, opens "Hardscrabble" mines.
1827	Lucius Langworthy, another brother and two sisters arrive in Galena from Quincy, Illinois.
June 1830	Sauk and Fox tribes abandon local mines after a Sioux attack in Dubuque area.
June 1830	Langworthy party crosses Mississippi to exploit deserted Indian mines; visits Dubuque's grave site.
July 4, 1830	Captain Zachary Taylor and a Prairie du Chien military detachment evict Langworthys and other miners from Dubuque area mines.
Sep. 21, 1832	Treaty ending the Black Hawk War signed.
Jan. 25, 1833	U.S. troops continue to evict lead prospectors from eastern Iowa since Black Hawk treaty is not yet ratified by the Senate.
June 1, 1833	Black Hawk cession now open to legal white settlement; government issues written permits to miners, licenses to smelters.
May 28, 1834	Miners vote to name their river settlement "Dubuque" at a public meeting.
1835	Dubuque now processing and exporting fifty million pounds of lead.
May 11, 1836	The *Du Buque Visitor* becomes city" first newspaper. Its motto: "Truth our guide—the public good our aim."
Dec. 28, 1846	Iowa admitted to the Union as the twenty-ninth state.

Dec. 23, 1854	George Rogers opens his gunsmith business—"declares war" on other local gun dealers in Dubuque.
Mar. 3, 1855	City of Dubuque transfers Mississippi riverfront property to James Langworthy for improvement work.
Mar. 7, 1855	Dubuque Harbor Improvement Company is incorporated.
Dec. 3, 1855	Newspapers report George W. Rogers intends to erect Shot Tower.
Feb. 13, 1856	George Rogers & Co. takes out mortgage with DHIC, with Shot Tower and planned improvements as collateral.
Feb. 28, 1856	L.H. Langworthy brings Jamestown, Wisconsin mining company representatives to Dubuque to discuss deep mining technology; calls for creation of a stock company to finance new mining methods.
Apr. 25, 1856	City of Dubuque transfers property to the Dubuque Harbor Improvement Company.
Aug. 20, 1856	*Daily Republican* reports Shot Tower construction underway; expected completion date: October 1, 1856.
Sep. 27, 1856	Reports say Shot Tower to be 150 feet tall.
Oct. 10, 1856	Dimensions of the Shot Tower revealed: 150 feet tall, 25 feet square at the base and 10 feet square at the top. 105 feet composed of limestone, 45 feet brick. Total investment: "$12,000."
Nov. 12, 1856	George Rogers & Co. Shot Tower ready for business.
Nov. 17, 1856	Shot Tower is now producing shot.
Nov. 18, 1856	Experimental "first shot" made at tower.
Nov. 19, 1856	Local paper details shot production during onsite visit.
Nov. 24, 1856	Rogers ready for shot production.
Nov. 26, 1856	Shot Tower described as 150 feet tall, with nine stories, 35 windows and a cost of nearly $7,000. Added outbuilding construction pushes cost to "$10,000."
Dec. 17, 1856	George Rogers & Co. advertises "patent shot" in the *Dubuque Express & Herald.*
Dec. 24, 1856	Rogers now filling orders for shot "at nearly St. Louis prices."
1857	Dubuque shot reaching Chicago markets; Shot Tower running at full production.
Mar. 7, 1857	Dubuque Harbor Improvement Company acquires property from Langworthy, et. al.
May 4, 1857	Dubuque City Council approves subdivision of riverfront property.

Sep. 22, 1857	*Daily Northwest* describes shot production from Shot Tower 150 feet tall.
Sep. 22, 1857	"Messrs. Cook and Langworthy" now own the Shot Tower.
Jan. 27, 1858	Riverfront property subdivided and recorded. Shot Tower land parcel designated Lot 24 of Block 7.
Mar. 26, 1859	Messrs. Tallman and Clark test new machinery in tower, anticipate manufacturing soon—the only written account naming these partners.
Apr. 14, 1859	Shot production supervised by Tallman and Clark; reports describe "130-ft." Shot Tower; W.H. Daniels is the only identified Shot Tower employee.
Feb. 13, 1860	*Express & Herald* advertizes marshal's auction of the George Rogers & Co. Shot Tower on Lot 24 of Block 7.
Mar. 14, 1860	Gilbert Woodruff pays $1,653 for the Shot Tower (Lot 24 of Block 7) at auction.
Mar. 15, 1860	Woodruff sells Shot Tower to J.K. Graves for $1,200.
Apr. 12, 1861	Civil War begins when Confederates fire on Fort Sumter in South Carolina.
Apr. 24, 1861	Flooding halts shot production for nearly two months.
May 24, 1861	Graves resumes shot production when flooding subsides. To celebrate anticipated Union victories, he purchases large American flag to fly atop the tower.
May 1861	J.K. Graves offers Iowa governor Kirkwood his Shot Tower's entire production for the war effort. Kirkwood never responds.
May 30, 1861	Height of American flag flown atop the Shot Tower at 160 feet—reported as the highest-flying flag in Iowa.
Sep. 12, 1861	76 kegs of shot shipped by rail from Dunleith, Illinois. Shot Tower owners contribute product display at annual County Ag. Fair.
Jan. 9, 1862	Weight of stored shot collapses tower finishing room's main floor; one employee injured. Production going well.
Mar. 19, 1862	Tower at full production, making 100–150 sacks of shot daily. Chicago firm contracts for 1,000 sacks per week in 1862. "Orders from a distance are pouring in more rapidly than they can possibly be filled."
July 13, 1862	Local boy steals arsenic chunk from Shot Tower, tosses it in family backyard. Family's chickens later discovered "dead upon the ground, and as many more on the point of giving up the ghost," indicating the structure is now idle and unattended.

July 14, 1862	J.K. Graves sells Shot Tower to St. Louis shot-making firm Chadbourne & Forster. *Daily Times* reports the $3,000 sale on July 18, 1862.
July 21, 1862	*Daily Herald* reports "a party of men" inspect a 130-foot mineral shaft; intend to make shot there.
July 27, 1862	Graves now producing shot in 140-foot abandoned mine shaft: "The shot averages considerably better than that made at the Tower."
Aug. 1, 1862	*Daily Times* reports "experiments are daily being tried" to make shot using mineral shafts: "There is no necessity to building $15,000 towers when a hole in the ground with an expenditure of $500 will do as well."
Aug. 3, 1862	*Daily Times* reports "Messrs. Booth, Carter, Chaney and J.K. Graves commence the erection tomorrow of a new two-story brick building over the shaft just west of Gen. Booth's residence" to make shot.
Aug. 26, 1871	The "old and conspicuous land mark," the Dubuque Shot Tower, is sold by Chadbourne & Forster for delinquent taxes.
Feb. 1, 1874	Equestrian oak statue of Andrew Jackson is reported atop the Shot Tower. Local attorney John Deery raises funds to hire steeplejack John Dreyhouse to install Thomas Kavanaugh's wooden sculpture. No reasons are given for this prank.
May 11, 1881	Dubuque businessman C.J. Lesure purchases Shot Tower for $425.
May 16, 1881	C.J. Lesure purchases additional property—Lots 22, 23 and 24 of Block 7—from Chadbourne & Forster for $425.
June 9, 1881	W.H. Day of the Empire Lumber Co. of Eau Claire, Wisconsin, and C.J. Lesure of Dubuque incorporate as the Standard Lumber Co., with $300,000 in capital stock.
June 24, 1881	Lesure sells property, including Shot Tower in Lots 22, 23 and 24 of Block 7, to Standard Lumber Co. for $600.
Oct. 31, 1897	Julien Dubuque reburied atop Catfish Creek overlook in a limestone monument designed by Dubuque Civil War artist Alexander Simplot.
May 26, 1911	Late-night fires of suspicious origin are set in Standard's lumberyard.
May 27, 1911	Lumber fire destroys Shot Tower's watchtower and internal wooden structure.

July 25, 1911	Standard Lumber Co. sells burned-out yards and heavily damaged Shot Tower to M.H. McCarthy for one dollar.
Oct. 23, 1915	Allison-Henderson Memorial Association purchases Shot Tower and surrounding property for $12,000, with future plans to honor Senator William Allison and Speaker of the House David Henderson.
Aug. 16, 1916	Allison-Henderson Memorial Association transfers Shot Tower and surrounding property to the City of Dubuque.
Oct. 11, 1931	Dubuque Women's Club petitions city to preserve tower and create a small park around it. City eager to develop area for industrial use.
Aug. 6, 1933	*Dubuque Telegraph Herald* celebrates the Shot Tower's seventy-fifth birthday, even though it is actually seventy-seven years old.
1934	Historic American Buildings Survey (HABS) program documents the Shot Tower with scale drawings and photos.
Jan. 2, 1949	*Des Moines Register* calls the Dubuque Shot Tower a "long-neglected historical monument."
Dec. 1, 1949	Shot Tower elevated to icon status by Dubuque historian William Peterson in Iowa State Historical Society's monthly journal, *The Palimpsest*.
Dec. 25, 1949	Dubuque Girl Scouts offer to plant shrubbery around Shot Tower if city agrees to clean up area. No action is taken.
Sep. 9, 1951	Engineer taking readings for industrial building foundations constructed north of the Shot Tower pronounces the structure sound and able to last "another 100 years."
Nov. 21, 1952	Dubuque Chamber of Commerce declares support for saving tower because of its "great historical interest" and its "trademark" role for the city of Dubuque.
Feb. 11, 1953	Dubuque Beautification Committee recommends sealing tower's lower openings, installing a roof and completing an engineering survey.
July 26, 1959	County historical society urges city council to proceed with Shot Tower restoration.
Oct. 25, 1959	County historical society joins with the Dubuque Vacation Committee to push for tower restoration.
Mar. 22, 1960	"Save Our Shot Tower" campaign raises $6,848.
Oct. 17, 1976	Dubuque Shot Tower is listed on the National Register of Historic Places by local historian Helen Mercer.

Oct. 17, 1976	Brewer Joseph Pickett and Davenport surgeon Richard Vermeer make shot in Shot Tower. Bottled shot is sold by the Dubuque County Historical Society to celebrate American bicentennial.
Sep. 28, 1987	*Telegraph Herald* reports serious consideration given to disassembling the Shot Tower and rebuilding it at the Ice Harbor for easier tourist accessibility.
Nov. 17, 1991	Runaway railroad car slams into Shot Tower, causes \$5–10,000 damage.
Apr. 12, 1992	*Telegraph Herald* telephone survey reveals 76 percent of Dubuque residents favor tower's preservation.
Jan. 16, 1993	Dubuque Shot Tower designated a municipal historic landmark.
2001	IIW Engineers & Surveyors of Dubuque conduct structural analysis of the Shot Tower.
2003	State Historic Preservation Office awards City of Dubuque a \$10,000 Historic Sites Preservation grant with top ranking.
2004	Save America's Treasures (SAT) grant for \$300,000 is submitted by city.
Oct. 2004	City is awarded \$295,000 from Save America's Treasures.
Jan. 18, 2005	Central Alternative High School students begin Shot Tower research project. Student Sean Bosveld discovers his ancestor, J.K. Graves, owned the tower.
May 2, 2005	City council approves Chicago firm of Hasbrouck Peterson Zimoch Sirirattumrong for tower rehabilitation consultants.
June 16, 2005	HPZS consultant inspects tower site.
Aug. 8, 2005	HPZS releases their study.
Jan. 17, 2006	Dubuque City Council proclaims 2006 as "The Year of the Shot Tower."
July 26, 2006	National Park Service conducts Shot Tower site visit.
Aug. 24, 2006	Geomorphological and archaeological investigation of tower site completed.
Oct. 9, 2006	Excavation and sorting of materials begins. Archaeologist Mike Gregory and Loras College archaeology students uncover tower's original well.
June 4, 2007	Dubuque City Council approves amending archaeological contract to remove the remainder of the fill in tower's interior.
July 18, 2007	Interior fill removed from tower. Consultants discover well deeper than indicated in 1934 HABS report. Excavated material used in constructing Star Brewery Drive.
May 15, 2008	Final archeological investigative report is submitted.

Aug. 3, 2009	City council rejects all bids for Shot Tower rehabilitation. Project reviewed to reduce costs.
Aug. 24, 2009	National Park Service approves grant extension to May 31, 2010, due to bid-letting complications.
Oct. 19, 2009	Bidding process begins for Historic Shot Tower Rehabilitation Project—Phase I Revised Scope.
Nov. 11, 2009	City of Dubuque receives grant reimbursement of $26,000 from Save America's Treasures.
Nov. 16, 2009	City council rejects all bids, authorizes rehabilitation project to be re-bid.
Dec. 21, 2009	City council awards rehabilitation contract.
Feb. 1, 2010	Work begins on Phase I of the Shot Tower rehabilitation project.
Apr. 14, 2010	Jackson Tuckpointing of Dubuque completes work on the Shot Tower.
June 17, 2010	Window replacement begins in tower.
July 20, 2010	Window installation and all exterior work completed. Tower interior to be excavated, with wood observation deck yet to be constructed.
Aug. 17, 2010	Phase I of the Shot Tower rehabilitation is completed.
Sep. 15, 2010	Project completed and all grant reimbursements are received. Grants are closed.

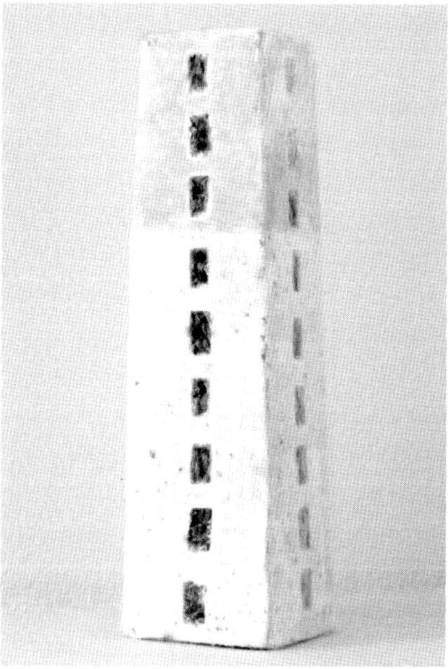

Behold the power of cheese! The famous pepper jack and cheddar cheese Shot Tower centerpiece created for the luncheon hosted by the Central Shot Tower Investigators when they kicked off their research project in January 2005. *John Adelmann.*

About the Editor

From 1975 through 2010, John Adelmann taught social studies at Central Alternative High School in Dubuque, Iowa. During his career, he used the community as a teaching partner, guiding his students into in-depth historical research with local connections. These projects generated topical research books written by the students and culminated with memorable public seminars that gave them and their audiences the opportunity to make personal relationships with celebrated individuals from history, including Gloria Owens Hemphill, daughter of 1936 Olympic gold medalist Jesse Owens; B-29 "Enola Gay" pilot Brigadier General Paul Tibbets; members of the renowned Tuskegee Airmen; Francis Gary Powers Jr., son of the U-2 spy pilot; noted Berlin Airlift pilot Colonel Gail Halvorsen; members of the famed "Little Rock Nine" civil rights pioneers; SR-71 reconnaissance aircraft engineer/designer William Fox; and Project Mercury astronaut Commander M. Scott Carpenter. Proceeds from sales of various student publications were contributed to the Commemorative Air Force's Red Tail Project in Minneapolis to help restore a P-51 Mustang aircraft to honor the Tuskegee Airmen, to the National World War II Veterans Memorial in Washington, D.C., and to help purchase books for school district libraries. This book's royalties will benefit the Center for Dubuque History and the Dubuque County Historical Society.

John's work has been recognized by the National Council for the Social Studies, the American Association of State and Local History, the State Historical Society of Iowa, the Iowa Historical Preservation Alliance, the Aerospace Education Foundation, the American Airpower Heritage Foundation, the International City/County Management Association, the Veterans of Foreign Wars and ING's "Unsung Heroes." In 2002, he became the first high school social studies educator in the tri-state area to be certified by the National Board for Professional Teaching Standards.